SpringerBriefs in Electrical and Computer Engineering

Series Editors

Woon-Seng Gan, School of Electrical and Electronic Engineering, Nanyang Technological University, Singapore, Singapore

C.-C. Jay Kuo, University of Southern California, Los Angeles, CA, USA

Thomas Fang Zheng, Research Institute of Information Technology, Tsinghua University, Beijing, China

Mauro Barni, Department of Information Engineering and Mathematics, University of Siena, Siena, Italy

SpringerBriefs present concise summaries of cutting-edge research and practical applications across a wide spectrum of fields. Featuring compact volumes of 50 to 125 pages, the series covers a range of content from professional to academic. Typical topics might include: timely report of state-of-the art analytical techniques, a bridge between new research results, as published in journal articles, and a contextual literature review, a snapshot of a hot or emerging topic, an in-depth case study or clinical example and a presentation of core concepts that students must understand in order to make independent contributions.

More information about this series at http://www.springer.com/series/10059

Lei Jiao

Channel Aggregation and Fragmentation for Traffic Flows

 Springer

Lei Jiao
The Department of Information
and Communication Technology
University of Agder
Grimstad, Norway

ISSN 2191-8112   ISSN 2191-8120 (electronic)
SpringerBriefs in Electrical and Computer Engineering
ISBN 978-3-030-33079-8   ISBN 978-3-030-33080-4 (eBook)
https://doi.org/10.1007/978-3-030-33080-4

This Springer imprint is published by the registered company Springer Nature Switzerland AG.
The registered company address is: Gewerbestrasse 11, 6330 Cham, Switzerland

Preface

Traditionally, a traffic flow is transmitted on a single channel. However, for the last decade, we have witnessed a fast improvement in the flexibility of channel utilization. Indeed, with the assistance of the channel aggregation (CA) and channel fragmentation (CF) techniques, more and more communication systems are able to support traffic flow transmissions on multiple channels or on a segmentation of a channel.

In this book, the impact of CA and CF on traffic flows is studied via analytical models, computer simulations, and test-bed implementations. In Chap. 1, we introduce the concept of CA and CF and summarize the existing research articles and communication protocols that are related to them. In Chap. 2, we review the basic concept and calculation of Markov chains, which are used as the analytical tool for modeling the communication systems in this book. Chapter 3 elaborates the modeling process of the system with CA and CF in the simplest scenario, where there is one type of users with one type of flows in the system. Chapter 4 presents the analytical process in a cognitive radio network, which is a typical system that has multiple types of users. Besides, in Chap. 4, simulation approaches are also explained in detail. To study the behavior of a real and operational system with CA and CF, Chap. 5 presents a test-bed implementation of such a system.

The book can serve as a reference for graduate students who are interested in CA and CF techniques, and who aim at obtaining better comprehension of the performance benefits when such techniques are employed for traffic flow transmission. The book can also be interesting to students who would like to learn the modeling process via Markov chains for performance evaluations. Indeed, the book provides a step-by-step guidance for Markov chain modeling and simulation, which does not only apply to the communication systems introduced in this book, but can also be used as a powerful analytical tool in other areas.

Grimstad, Norway   Lei Jiao
August 2019

Acknowledgements

I would like to express my sincere thanks to my institute, the University of Agder, for providing me with a healthy working environment. Most of the research work involved in this book has been carried out in the University of Agder.

For technical discussions and suggestions, I am grateful to Dr. Indika Anuradha Mendis Balapuwaduge. I am also thankful to Miss Maria Thalina Broen for her careful proofreading and editing on the manuscript. I would also like to record my thanks to Dr. Xuan Zhang, for stimulating discussions, advice on writing strategies, and help on various ways to make this book more readable.

Last but not least, my loving thanks go to my two adorable daughters. To the 3-year-old Miriam for keeping away from my keyboard when I was working from home. To the 5-year-old big sister, Eva, for having secretly pressed my keyboard a couple of times and messed up the entire file so that it refused to compile. Both have the potential to become proper engineers in the future, in my opinion.

Contents

Abbreviations

3GPP	3rd Generation Partnership Project
BDP	Birth and Death Process
CA	Channel Aggregation
CC	Component Carriers
CCA	Constant Channel Aggregation
CF	Channel Fragmentation
CR	Cognitive Radio
CRNs	Cognitive Radio Networks
CTMC	Continuous Time Markov Chain
DFA	Dynamic Fully Adjustable Strategy
DPA	Dynamic Partially Adjustable Strategy
Dy	Dynamic Strategy
EFAFS	Extended Full Adaptation and Full Sharing Strategy
FCC	Federal Communications Commission
MAC	Medium Access Control
MC	Markov Chain
NI	National Instruments
OFDMA	Orthogonal Frequency Division Multiple Access
OSA	Opportunistic Spectrum Access
PUs	Primary Users
P-VCA	Probability Distribution based Variable Channel Aggregation
QoS	Quality of Service
QSR	Quasi-Stationary Regime
R-VCA	Residual Channel based Variable Channel Aggregation
Rx	Receiver
SDR	Software Defined Radio
SS	Spectrum Sharing
SUs	Secondary Users
TCP	Transmission Control Protocol
Tx	Transmitter
UDP	User Datagram Protocol

USRP	Universal Software Radio Peripheral
WFQ	Weighted Fair Queuing
WLANs	Wireless Local Area Networks
WSNs	Wireless Sensor Networks

List of Figures

List of Tables

Chapter 1
Introduction

With the rapid development of modern communication systems and electronics technologies, spectrum utilization becomes more and more flexible and dynamic. Traditionally, a traffic flow is sent within one communication channel. With the help of channel aggregation (CA) technology, it is possible to adopt multiple channels for transmitting one flow, while the channel fragmentation (CF) technology can help divide one channel into multiple segments in order to transmit multiple flows. Studies on CA and CF and their relevant topics are numerous. To indicate the amount of the studies, we searched *channel aggregation* as the keyword in IEEE *Xplore*, on January 20th, 2019, and found 1256 relevant articles. In this chapter, we introduce the principle of CA and CF, and the concepts that are similar to them. We also provide an incomplete survey of these techniques with the main focus on cognitive radio networks[1] (CRNs) [1].

1.1 Channel Aggregation and Related Concepts

CA [2, 3] is a technique that combines multiple channels for a single communication event, instead of using merely one channel all the time. With the CA technique, communication pairs could take advantages of adopting several neighboring channels, as well as the separated ones in the frequency domain, and utilize them simultaneously for transmissions as if they are one integrated channel. There

---

[1]Cognitive radio (CR) is a software-defined radio with built-in intelligence that can detect available spectrum opportunities in a wireless spectrum and adaptively adjust transmission parameters to operate concurrently with existing systems. A CR normally does not own a communication spectrum and thus needs to access channels belonging to other systems opportunistically. A CRN is a communication network composed by CRs.

© The Author(s), under exclusive license to Springer Nature Switzerland AG 2020
L. Jiao, *Channel Aggregation and Fragmentation for Traffic Flows*,
SpringerBriefs in Electrical and Computer Engineering,
https://doi.org/10.1007/978-3-030-33080-4_1

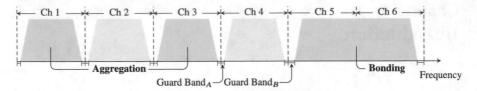

Fig. 1.1 Illustration of channel bonding and channel aggregation

are several terminologies and concepts that are similar to CA, such as channel bonding [3, 4], carrier aggregation [5, 6], and channel assembling [7, 8]. We will explain these concepts via the example shown in Fig. 1.1.

Figure 1.1 illustrates channel occupation at a particular time instant when two channels can be utilized at the same time by one communication pair. Communication pairs combine two channels in two different ways. One is *aggregating* two separated idle channels, like Channels 1 and 3, into one channel; the other is to *bond* two neighboring idle channels, for instance, Channels 5 and 6, into one. Note that neighboring channels can also be *aggregated* with the guard band being kept in between. The key feature that differentiates CA from channel bonding is that the former retains the channel's identity in the spectrum domain [2]. In other words, CA keeps the original channel structure in the spectrum allocation with a guard band being applied between two neighboring channels, whereas channel bonding merges adjacent spectrum as one channel, with the guard band between the original neighboring channels being also integrated into it. Clearly, if the vacant channels are neighboring to each other, channel bonding is able to utilize the integrated guard bands for data transmission, although a larger guard band is required at the edges of the bonded channel, as shown by Guard Band$_B$ in the same figure [3]. It is shown that when channels are neighboring to each other, channel bonding, compared with CA, is with less additional overhead and less complexity [2, 4].

The term channel *assembling* is utilized to refer to channel bonding and CA collectively [4, 7, 8]. It is mostly used in scenarios where there are multiple channels that have been utilized for one communication event, and where we do not need to differentiate the physical layer details.

Carrier aggregation is similar to channel aggregation, and is mostly utilized in the releases from the 3rd generation partnership project (3GPP) [5], and the research articles in the same direction [6, 9–11]. The concept of carrier aggregation and its terms and ways of operation are defined in detail in [5].

The core of the concept of CA, channel bonding, carrier aggregation, and channel assembling, is to apply multiple frequency resources to one communication event. As mentioned before, this concept has been extensively studied in the last 10 years via proposed medium access control (MAC) protocols and dynamic spectrum access strategies [7, 12–14] for improving system performance. In addition, this technique has also been applied in many recent communication standards, such as LTE-A [15], the 802.11 family (e.g. 802.11n/ac) [16], 802.22 [17], and 5G [18].

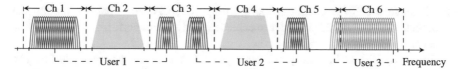

Fig. 1.2   Illustration of channel fragmentation together with channel aggregation

1.2   Channel Fragmentation Together with Channel Aggregation

As opposed to CA that allows a communication pair to use multiple channels at the same time, CF [19] is a technique that enables a communication pair to access a *portion* of a channel for transmission purpose. The concept is similar to multiplexing in the frequency domain, and is particularly useful for cognitive radios (CRs) when the spectrum band is significantly larger than the spectrum that one CR requires [19]. In recent research and practices, CF has very often been enabled together with CA [7, 20, 21]. This is because, with CA enabled along with CF, a communication pair is able to utilize not only a fragment of a channel but also multiple channel fragments. Figure 1.2 illustrates an example of CF enabled with CA, where User 1 utilizes Channel 1 and a fragment of Channel 3, User 2 uses fragments from Channel 3 and Channel 5, and User 3 has the whole band of Channel 6 and a small portion of Channel 5. We thus can say that CF, together with CA, provides more flexibility than traditional schemes in terms of channel access. The technique that supports CF and CA at physical layer is orthogonal frequency division multiple access (OFDMA) [22]. OFDMA allocates various subcarriers to distinct users, and provides the finest granularity that the system allows for resource allocation.

1.3   Channel Access Schemes with Channel Aggregation and Fragmentation

In this section, we briefly recap the existing communication standards that are applying CA and CF, from different standardization bodies. We also roughly survey research articles on CA and CF from academia, with the main focus on CRNs.

1.3.1   Communication Standards Applying CA and CF

In communication industry, many recent standards support CA or its similar concept for medium access, such as IEEE 802.22 [17, 23], 802.11ac [16], LTE-A [15], and

5G [18]. In the IEEE 802.22 working group, CF, CA, and channel bonding are specified and studied [23] in order to achieve good system performance and flexible spectrum utilization.

In the IEEE 802.11 family, the concept of CA is firstly presented in 802.11n, where communications can employ channels with bandwidths 20 and 40 MHz. In IEEE 802.11ac [16], it is standardized that the system can use 20, 40, 80, 160 MHz for communication purpose, depending on different communication configurations and environments. This standard provides not only more flexibility of spectrum utilization but also significant increment in physical-layer data rate, compared with legacy standards from the same family, such as 802.11a/b.

In the standards from 3GPP, carrier aggregation is applied firstly in LTE-A and subsequently in 5G. In Release 10 of the 3GPP specifications, carrier aggregation is standardized to allow a collection of up to five LTE component carriers (CCs) with the system bandwidth up to 100 MHz (5×20 MHz). In Release 11, they established the performance specifications for new inter-band and intra-band combinations and defined the core requirements for intra-band non-contiguous deployments as well. The technique of carrier aggregation is further enhanced in various aspects through Release 12, with the maximum number of carriers that can be aggregated being kept the same as that in the previous releases though. Then in Release 13, the number of 20 MHz CCs that can be aggregated is increased from 5 to 32, which, in principle, enables LTE terminals to apply a bandwidth up to 640 MHz, most of which will be located in the unlicensed spectrum in the 5 GHz band.

The operation of carrier aggregation defined in 3GPP is relatively flexible. A mobile terminal may utilize one or more CCs depending on its own capabilities. The number of CCs that can be aggregated in the uplink can be different from that in the downlink, with the constraint that the number in the uplink is not greater than the number in the downlink. Besides, the aggregated CCs may belong to the same band, which is called intra-band carrier aggregation, within which, contiguous carrier aggregation is to aggregate contiguous CCs, whereas non-contiguous carrier aggregation aggregates non-contiguous CCs. The aggregated CCs can also belong to different spectrum bands, and this is what we call inter-band carrier aggregation. The inter-band carrier aggregation is more complex and requires more advanced transceivers. Figure 1.3 illustrates the CCs in the above three carrier aggregation cases, namely, intra-band continuous carrier aggregation, intra-band non-continuous carrier aggregation, and inter-band carrier aggregation.

1.3.2　MAC Protocols in Research Articles Employing CA and CF

In the academic society, there are tremendous research activities conducted on CA and CF, many of which have applied these techniques in improving performance

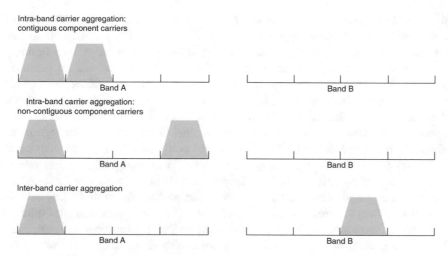

Fig. 1.3 Basic concept of carrier aggregation in LTE-A

of traditional wireless networks such as wireless sensor networks[2] (WSNs) [24], wireless local area networks (WLAN) [25–31], and CRNs [17, 32, 33]. One can refer to [4, 34, 35] for comprehensive surveys on these techniques being applied in different communication systems. In the following paragraphs, we focus on the research articles that employ CF and CA in CRNs. The main reason for emphasizing CRNs is because its operation concept is much more complicated than that of the other communication systems. This is due to the random behavior of primary users[3] (PUs) and the opportunistic channel access behavior of CRs. Because of the stochastic nature of CRNs, both CA and CF are considered to be fitting and applicable to the system in order to enhance the performance.

The concept of CA and CF has been widely employed in existing MAC protocols for CRNs, in both centralized protocols [17, 32] and the distributed ones [12, 13, 36–39]. In centralized MAC protocols, such as the one defined in IEEE 802.22 and the one proposed in [32], there exists a central controller that coordinates various users with different services in channel accessing. However, in distributed schemes, there are no such centralized coordinators, and thus secondary users[4] (SUs) usually compete for channel access by following a certain protocol.

In many distributed MAC protocols for CRNs, when SUs compete with each other for channel access, only the SU that wins the competition can utilize

---

[2]Wireless sensor networks are composed by a number of dispersed sensors with communication modules for monitoring or collecting data from a specific environment. WSNs have been applied in many fields, such as industrial and home automation.

[3]Primary users are the owners of the spectrum who have priority in spectrum access in the CRNs.

[4]Secondary users in CRNs are users who borrow the spectrum belonging to PUs. We use SUs and CRs interchangeably when there is no ambiguity.

the available channels, while the other SUs have to wait for the next round of competition to gain their opportunities [13, 37, 40]. In cases with time-slotted channels [13, 37], once an SU senses that there are no PU activities at the beginning of a slot on a channel, then for the rest of the slot, the channel will be vacant in terms of PU activities. In such cases, it is beneficial for the SU winner to assemble all the channels that are available in that time slot for transmission, as higher data rate can be achieved without interruptions from PUs. However, if PUs may appear at any time after an SU sensing the channel being vacant, more aggregated channels may result in a higher probability of collision with PUs. In such cases with unslotted channels, more aggregated channels provide a higher data rate at a higher risk of colliding with PUs. Therefore, SUs need to balance between the number of aggregated channels and the collision probability on these channels. In [40], a statistical channel allocation MAC protocol that operates on such unslotted channels is introduced. In that work, the scenario that SU transmissions terminate because of packet collisions with PUs is examined by simulations, and a strategy is also designed to avoid such collisions.

In the centralized MAC protocols where CA and CF are enabled in CRNs, it is often the case that a central controller schedules parallel SU services [17, 32, 36] by managing a control channel or a time segment of a frame dedicated for allocating channels. The scheduling in the centralized solutions is challenging. For example, new SU arrivals may have limited resources or even be blocked from communication if ongoing SU services occupy too many channels. However, if too many channels are reserved for new arrivals, the resource may end up being wasted when there are not as many SU arrivals as expected. Both these cases degrade the system's performance, and hence a properly designed scheduling scheme is desired to avoid both of them. In addition, when multiple types of flows exist in the system, especially when flows have various quality of service[5] (QoS) requirements or priorities, a sophisticated scheduling scheme is necessary for maintaining system performance.

1.4 Mathematical Analyses for CA and CF in CRNs

Although many standards and protocols have already employed CA and CF, mathematical studies on the CA and CF techniques are still of great interests, as they reveal the nature and the property of the associated strategies. The mathematical tools that have been most used in the study of CA and CF in CRNs can be categorized, in a broad sense, into two groups. One group includes the study of

---

[5]Quality of service describes the overall performance of a service in a communications system, which is usually quantitatively measured via parameters such as throughput, delay, jitter, packet loss ratio, and etc.

resource allocation approaches[6] with the main scope on physical layer and MAC layer, and optimization theory is the most adopted mathematical tool [41]. The other group mainly deals with packets or traffic flows, focusing on the design of scheduling strategies and the evaluation of their performance. Markov chain (MC) and queuing theory are most utilized as the mathematical tools within this group [42, 43]. In this section, we briefly summarize the studies on these two groups.

1.4.1 Resource Allocation Approaches in CRNs with CA and CF

In CRNs, there are two ways of spectrum reuse: spectrum sharing (SS) [44–46] and opportunistic spectrum access (OSA) [12, 13, 17, 47]. In the SS mode, the same band can be utilized by PUs and SUs simultaneously as long as the interference with the PUs that is caused by the SUs is below a certain threshold [44, 48]. A metric for the interference level, i.e., the interference temperature, is proposed by the federal communications commission (FCC) for interference analysis [49]. Although the concept of SS is utilized in many research papers, it is in practice difficult to apply. Indeed, FCC has abandoned the rulemaking for interference temperature as no parties provided information on specific technical rules that could be adopted to implement it [50]. In the OSA mode, the SUs can transmit over a band only if PUs do not utilize it. The decision whether SUs can transmit or not is made partially according to the result of spectrum sensing that aims to detect the presence of PU activities. If the sensing indicates no PU activities on the current channel, SUs can access the channel for transmission. When a PU service re-appears to a channel that is in use by SUs, the SUs have to stop transmission and release the channel immediately. Note that in both SS and OSA, SUs should not interfere with PUs. In this book, we focus only on the OSA mode due to its applicability over licensed spectrum.

In many channel access schemes for OSA, it is usually assumed that the PU activities are time-slotted [13, 37, 51]. With this assumption, once a channel is sensed to be unoccupied by PUs, it will remain free from PU activities for the rest of the time slot. Therefore, in between sensing intervals, the collision of ongoing SU services with PU activities will not happen and the idle channels can be utilized by SU packet transmissions without considering the re-appearance of PUs. In this case, the traditional water-filling algorithm [52] can be applied to the aggregated multiple idle channels when the data rate is to be maximized given a limited power budget. Similarly, when the time scale for the PU activities greatly exceeds that of the SUs' [12, 17, 47], the collisions between PUs and SUs are ignorable, and therefore the available channels can be deemed as dedicated to SUs. In this case,

---

[6]We consider channel selection as a special case of resource allocation. If no power is allocated on a channel, the channel is not opted for communication.

static spectrum allocation policies and the traditional water-filling algorithm can be considered as the optimal solutions for resource allocation because there are no considerable changes in the available spectrum.

When PUs are relatively active and may appear at any time on the channels, resource allocation becomes more complicated. This is because collisions due to re-appearance of PU activities during an SU transmission need to be considered. If such a collision happens, SUs need to vacate the channel immediately to give way to PUs, resulting in possible packet loss for SUs [40, 53]. Indeed, in order for the SUs to utilize the spectrum efficiently, it is not only important for them to maximize the data rate by aggregating multiple channels but also critical that they can keep the collisions with PUs below a certain level. This means a decision needs to be made to determine how many and which channels are to be aggregated for an SU's transmission, and the decision reflects the SUs' ability to select the best available spectrum in order to satisfy their QoS requirements without causing harmful interference with the PUs. The spectrum decision usually involves the statistical analysis on channel conditions and the modeling of PU activities, which are commonly based on historical data that are obtained from spectrum sensing or the knowledge learned from a third party as a priori. For example, in [40], a statistical channel selection scheme is introduced in their proposed MAC protocol, and therein channels for packet transmissions are selected based on the statistical characteristics of PU events on different channels. Power allocation on those channels is neither analyzed nor optimized in [40] though.

To optimize the spectrum opportunities in the dynamic environment where PUs are active, the resource allocation process is usually formulated as an optimization problem. After the problem itself is defined, algorithms are designed to solve the formulated problem. For instance, in [33], when there are multiple idle channels exist in the OSA system, the optimization problem is formulated to maximize the data rate of SUs under the constraint that the collision probability with the PUs must be limited. The problem is shown to be NP-hard and an algorithm is then proposed for a sub-optimal solution in polynomial time.

Due to the randomness of the PUs' behavior, the formulated problems in CRNs with CA and CF are usually more complicated than those in traditional scenarios. Algorithms that provide solutions to these problems form the foundations based on which upper layer services, such as traffic flows, are able to explore further possibilities in improving the performance of the system.

1.4.2   MC Analysis for CRNs with CA and CF

As mentioned in Sect. 1.4.1, resource allocation in CRNs is usually carried out in physical layer and MAC layer, and optimization techniques are usually adopted as the mathematical tool for analyzing and improving system performance. With the help of these techniques, scheduling of traffic flows in the systems with CF and CA becomes possible. However, when it comes to performance analysis for different scheduling schemes in the packet level and flow level, MC based analysis is the mathematical method that is usually applied. In this subsection, we summarize the existing MC based multi-channel scheduling strategies for traffic flows in CRNs with CA and CF. While packet-level analysis requires physical layer details, the MC based analyses at flow level usually do not differentiate the physical layer details of CA and channel bonding, as long as the relationship between the number of channels and the service rate of the MC is clearly defined. Therefore, we use the term, CA, to represent the case when multiple channels are utilized for one flow and do not specify which physical technology is applied unless otherwise stated.

The performance of systems with CF, i.e., the ones where a PU channel can be split into several fragments for SU flows to transmit, is studied through continuous time MCs (CTMCs) [19, 54, 55] by examining parameters such as throughput, blocking probability, and forced termination probability. The system where there are infinite number of users is studied in [19, 54], with both handover and non-handover cases being analyzed. Handover means that ongoing SU flows can move to another vacant channel upon PU's re-appearance in the current channel. The authors [19, 54] also investigated the concept of channel reservation, which is a trade-off between throughput and blocking probability. In [55], the performance of the CRNs with a finite number of users is investigated in the quasi-stationary regime (QSR), where QSR describes a scenario where the time scale of PU's activity is significantly greater than that of SU's.[7] The studies for the schemes in [19, 54, 55] do not consider CA in the systems.

The strategies of CA can be sorted into two categories depending on whether spectrum adaptation is supported or not. When spectrum adaptation is not supported, the strategy is normally named as a static strategy. When spectrum adaptation is employed, the strategy is usually called a dynamic strategy. The meaning of spectrum adaptation is twofold [7]. On the one hand, similarly to spectrum handover, spectrum adaptation allows SUs to switch an ongoing flow to another vacant channel (if it exists), when a PU re-appears on the current channel. On the other hand, an ongoing SU flow can adapt the number of channels it uses according to PUs' and other SUs' activities. In other words, spectrum adaptation enables ongoing SU flows to change the number of channels they occupy adaptively during transmission, in addition to spectrum handover. However, if spectrum adaptation is not fully enabled, i.e., if it is enabled with spectrum handover only or even without spectrum handover, then ongoing SU flows will not be able to adjust the number of the channels they use. How spectrum adaptation is utilized depends on specific channel access strategies.

When spectrum adaptation is not fully supported, no matter whether spectrum handover is supported or not, CA strategies can hardly achieve the performance gain in terms of capacity, forced termination probability, and blocking probability in comparison with the traditional strategies without CA. The main reason is that when ongoing SU flows cannot adjust the number of the channels they utilize, they will suffer from high forced termination probability due to re-appearance of

---

[7]More detailed definition of QSR can be found in Sect. 4.1.3.

PUs, and they will also block newly arrived SU services, resulting in poor overall performance. These observations are obtained via MC analyses from different perspectives in [56] and [57]. In [56], three different CA strategies are examined with handover disabled, and they are the strategy without CA, the strategy with a fixed number of aggregated channels, and the strategy that adopts all idle channels when SU flows access the system. In [57], the so-called constant CA (CCA), probability distribution based variable CA (P-VCA), and residual channel based variable CA (R-VCA) are investigated. CCA is the same as the one with a fixed number of assembled channels in [56]. In P-VCA, the number of channels for SU flows follows a probability distribution, while the number of channels in R-VCA is determined based on the number of residual channels that are not occupied by PUs and SUs. Both the case with spectrum handover enabled and the one without spectrum handover are examined in [57]. A common feature of these strategies in [56] and [57] is that ongoing SU services are not able to change the number of channels, and thus performance gain is hardly observed for these strategies.

To overcome the drawbacks presented in [56, 57], CA strategies with fully enabled spectrum adaptation are proposed in [42, 58, 59]. Since the operation concept of CRNs is to sense the spectrum and access the channels opportunistically, fully enabled spectrum adaptation is essential. In [58], two CA strategies, i.e., the Greedy strategy and the Dynamic strategy, are proposed. In the Greedy strategy, ongoing SU flows have higher priority than the newly arrived ones. In other words, when an SU flow arrives, ongoing SU flows will not share their occupied channels with the newcomer if there are not enough idle channels at the moment for the new SU flow to commence. This results in high blocking probability of SU flows, which further leads to limited capacity gain of the Greedy strategy. The Dynamic strategy, on the contrary, allows ongoing SU flows to donate their occupied channels to the new coming SU flows, and its decent performance gain, in comparison with the strategy without CA, has been observed from different aspects through the numerical results. Motivated by the advantages brought by the ongoing SU flows' ability to donate their occupied channels to the newcomers, another dynamic strategy, i.e., the dynamic fully adjustable (DFA) strategy, is modeled and investigated in [59]. The main difference between the Dynamic strategy in [58] and the one in [59] lies in how SUs react upon a PU re-appearance. More specifically, the Dynamic strategy in [58] reduces the number of channels that an ongoing SU flow utilizes if PUs re-appear to one of these channels and if no idle channels exist, whereas DFA always requests the ongoing SU flows with the maximum number of channels to donate their channels to the PUs if there are not any idle channels. Clearly, DFA has higher flexibility, and thus better performance has been observed from DFA. The impact of queuing together with CA is studied in [8] for distinct flow types.

To study the benefit of CF for traffic flows, the authors of [7, 43], and [60] studied the performance of CRNs when both CA and CF are enabled with spectrum adaptation, via CTMC. Though the main goal for [43] is to show the capacity upper bound of CA strategies in the QSR, and the aim for [60] is to analyze the impact of guard band, they both examined the case where at least one channel is required by an SU flow to transmit, and showed that CF provides flexibility to the system in the sense that the system can employ a non-integer number of channels for a traffic flow. The case where an SU flow requires less than one channel for transmission is studied in [7], which evaluates the system performance for two different types of traffic flows that are supported by CF. The main results of [7] indicate that to fully exploit the advantage of CF, it is important to allow traffic flows to utilize less than one channel if the QoS criteria can be satisfied. Besides, the capacity upper bound of the strategies with CA and CF is derived in the QSR in [7].

To summarize, various CA and CF strategies for traffic flows have been evaluated under different conditions, and due to the stochastic nature of tele-traffic, MC is one of the most popular mathematical methods that are employed in analyzing the system performance for traffic flows. This book focuses on performance analyses based on MCs.

1.5   The Main Goal and the Organization of the Book

This book is to investigate the impact of CA and CF on system performance for traffic flows, by using MC as the tool for mathematical analysis. Its goal is to give readers a guided tour to apply MC in analyzing the performance of the system. It starts with the simplest case where there is only one type of users with one type of flows, then it moves on to a more complicated scenario where multiple types of users co-exist. The analytical method utilized in this book can be considered as an example of applying MC in doing generic modeling for stochastic process, and can also be employed to do performance analysis for other similar systems with stochastic features.

Indeed, there are many other aspects besides CA and CF, for example, queuing, that can influence the system performance for traffic flows. However, they are out of the scope of this book, as we only focus our study on the impact of CA and CF. To further make the analysis tangible, we ignore also the physical layer particularities, such as OFDM that can be adopted by resource allocation algorithms, by making a couple of assumptions[8] as follows:

1. There are certain schemes at both physical layer and MAC layer that can support CA and CF, so that traffic flows can be accommodated by a certain number of aggregated channels or by a fragment of a channel, according to the scheduling strategy at flow level.
2. The characteristics of a channel are random and stationary, and so is the statistics of the length of a certain type of flows. This guarantees that the statistical features of the service time for a certain type of flow on these channels will not change along time.
3. The arriving process of a certain type of traffic flows is random and stationary.

---

[8]For a complete system description, please refer to Sect. 3.1.2.

The remaining part of the book is organized as follows:

Chapter 2 reviews the basic theory of MC in discrete time and continuous time, which is the fundamental tool for analyzing different CA and CF strategies in this book. Readers who are already familiar with the concept and the derivation of stationary distribution of MC can skip this chapter.

Chapter 3 uses CTMC to study the impact of CA and CF at flow level on the simplest system where there exists only one type of users with one type of traffic flows. The behavior of various CF and CA strategies is modeled and the system performance is evaluated thoroughly. The capacity upper bound[9] for such a simple system with CF and CA is also derived in a closed form. This simple scenario allows readers to capture the essence of the problem in a relatively straightforward way. It also helps readers obtain fundamental comprehension on how to apply CTMC in analytical problems. Indeed, this chapter can be considered as the foundation for studying more complicated scenarios.

In Chap. 4, we study a more complicated scenario where there exist multiple types of users, each of which has one type of flows. Various CF and CA strategies are modeled and evaluated in a scenario of CRNs. In addition to the analytical models, simulation approaches are also presented in the same chapter.

Indeed, Chaps. 3 and 4 deal with analytical models of different CA and CF strategies in distinct scenarios, and are the main body of this book. The motivation for presenting these models is the popularity of CF and CA in the literature versus the lack of systematic and step-by-step interpretation of the modeling process for systems with these techniques at flow level. Such models can provide us with insights of these techniques and thus help us select the proper strategies in designing and implementing a communication system. Besides, the modeling procedure itself can be used as an example to study MC based analysis in communication systems.

We conclude the book in Chap. 5 by carrying out a test-bed study. The test-bed is based on an LTE architecture via a transceiver and its controller made by National Instrument. This test-bed study confirms the advantages of CA and CF in practice.

References

1. Mitola J, Maguire GQ (1999) Cognitive radio: making software radios more personal. IEEE Pers Commun 6(4):13–18. https://doi.org/10.1109/98.788210
2. Khalona R, Stanwood K (2006) Channel aggregation summary. IEEE 802.22 WG. https://mentor.ieee.org/802.22/dcn/06/22-06-0204-00-0000-channel-aggregation-summary.ppt
3. Cordeiro C, Ghosh M (2006) Channel bonding vs. channel aggregation: facts and faction. IEEE 802.22 WG. https://mentor.ieee.org/802.22/dcn/06/22-06-0108-00-0000-bonding-vs-aggregation.ppt
4. Bukhari SHR, Rehmani MH, Siraj S (2016) A survey of channel bonding for wireless networks and guidelines of channel bonding for futuristic cognitive radio sensor networks. IEEE Commun Surv Tutorials 18(2):924–948

---

[9]This capacity is different from Shannon capacity. More specifically, the capacity is defined by the number of completed flows over a unit time.

5. 3GPP (2017) 3rd Generation Partnership Project; Technical specification group radio access network; Evolved universal terrestrial radio access (E-UTRA); User equipment (UE) radio transmission and reception (release 12)
6. Cao Y, Sunde EJ, Chen K (2019) Multiplying channel capacity: aggregation of fragmented spectral resources. IEEE Microw Mag 20(1):70–77
7. Jiao L, Balapuwaduge IAM, Li FY, Pla V (2014) On the performance of channel assembling and fragmentation in cognitive radio networks. IEEE Trans Wirel Commun 13(10):5661–5675
8. Balapuwaduge IAM, Jiao L, Pla V, Li FY (2014) Channel assembling with priority-based queues in cognitive radio networks: strategies and performance evaluation. IEEE Trans Wirel Commun 13(2):630–645
9. Rostami S, Arshad K, Rapajic P (2018) Optimum radio resource management in carrier aggregation based LTE-advanced systems. IEEE Trans Veh Technol 67(1):580–589
10. Yuan G, Zhang X, Wang W, Yang Y (2010) Carrier aggregation for LTE-advanced mobile communication systems. IEEE Commun Mag 48(2):88–93
11. Maheshwari MK, Roy A, Saxena N (2019) DRX over LAA-LTE-a new design and analysis based on semi-Markov model. IEEE Trans Mobile Comput 18(2):276–289
12. Yuan Y, Bahl P, Chandra R, Chou PA, Ferrell JI, Moscibroda T, Narlanka S, Wu Y (2007) KNOWS: Cognitive radio networks over white spaces. In: Proceedings of IEEE DySPAN, Dublin, Ireland
13. Su H, Zhang X (2008) Cross-layer based opportunistic MAC protocols for QoS provisionings over cognitive radio mobile wireless networks. IEEE J Sel Areas Commun 26(1):118–129
14. Cao L, Yang L, Zheng H (2010) The impact of frequency-agility on dynamic spectrum sharing. In: Proceedings of IEEE DySPAN, Singapore
15. Parkvall S, Furuskar A, Dahlman E (2011) Evolution of LTE toward IMT-Advanced. IEEE Commun Mag 49(2):84–91
16. IEEE 802.11 task group ac (2013) IEEE Std 802.11ac-2013 (Amendment to IEEE Std 802.11-2012, as amended by IEEE Std 802.11ae-2012, IEEE Std 802.11aa 2012, and IEEE Std 802.11ad-2012) pp 1–425
17. IEEE 802.22 WG (2011) IEEE standard for wireless regional area networks part 22: Cognitive wireless RAN medium access control (MAC) and physical layer (PHY) specifications: policies and procedures for operation in the TV bands
18. Morgado A, Huq KMS, Mumtaz S, Rodriguez J (2018) A survey of 5G technologies: regulatory, standardization and industrial perspectives. Digital Commun Netw 4(2):87–97
19. Zhu X, Shen L, Yum TP (2007) Analysis of cognitive radio spectrum access with optimal channel reservation. IEEE Commun Lett 11(4):304–306
20. Anand S, Sengupta S, Hong K, Subbalakshmi KP, Chandramouli R, Cam H (2014) Exploiting channel fragmentation and aggregation/bonding to create security vulnerabilities. IEEE Trans Veh Technol 63(8):3867–3874
21. Coffman E, Robert P, Simatos F, Tarumi S, Zussman G (2012) A performance analysis of channel fragmentation in dynamic spectrum access systems. Queueing Syst 71(3):293–320
22. Holma H, Toskala A (2009) LTE for UMTS: OFDMA and SC-FDMA based radio access. Wiley, Chichester
23. Anand S, Hong K, Sengupta S, Chandramouli R (2011) Is channel fragmentation/bonding in IEEE 802.22 networks secure? In: Proceedings of IEEE ICC, pp 1–5
24. Ghods F, Yousefi H, Hemmatyar AMA, Movaghar A (2013) MC-MLAS: Multi-channel minimum latency aggregation scheduling in wireless sensor networks. Comput Netw 57(18):3812–3825
25. Joshi S, Pawelczak P, Cabric D, Villasenor J (2012) When channel bonding is beneficial for opportunistic spectrum access networks. IEEE Trans Wirel Commun 11(11):3942–3956
26. Lin Z, Ghosh M, Demir A (2013) A comparison of MAC aggregation vs. PHY bonding for WLANs in TV white spaces. In: Proceedings of IEEE PIMRC, pp 1829–1834
27. Chandra R, Mahajan R, Moscibroda T, Raghavendra R, Bahl P (2008) A case for adapting channel width in wireless networks. In: Proceedings of ACM SIGCOMM, New York, NY, USA, pp 135–146

28. Xu L, Yamamoto K, Yoshida S (2007) Performance comparison between channel-bonding and multi-channel CSMA. In: Proceedings of IEEE WCNC, pp 406–410
29. Deek L, Garcia-Villegas E, Belding E, Lee S, Almeroth K (2013) Joint rate and channel width adaptation for 802.11 MIMO wireless networks. In: Proceedings of IEEE SECON, pp 167–175
30. Deek L, Garcia-Villegas E, Belding E, Lee S, Almeroth K (2014) Intelligent channel bonding in 802.11n WLANs. IEEE Transactions on Mobile Computing 13(6):1242–1255
31. Wang X, Huang P, Xie J, Li M (2014) OFDMA-based channel-width adaptation in wireless mesh networks. IEEE Trans Veh Technol 63(8):4039–4052
32. Zhang W, Wang C, Ge X, Chen Y (2018) Enhanced 5G cognitive radio networks based on spectrum sharing and spectrum aggregation. IEEE Trans Commun 66(12):6304–6316
33. Jiao L, Razaviyayn M, Song E, Luo ZQ, Li FY (2011) Power allocation in multi-channel cognitive radio networks with channel assembling. In: Proceedings of IEEE SPAWC, pp 86–90
34. Ramaboli AL, Falowo OE, Chan AH (2012) Bandwidth aggregation in heterogeneous wireless networks: a survey of current approaches and issues. J Netw Comput Appl 35(6):1674–1690
35. Khan Z, Ahmadi H, Hossain E, Coupechoux M, Dasilva LA, Lehtomäki JJ (2014) Carrier aggregation/channel bonding in next generation cellular networks: methods and challenges. IEEE Netw 28(6):34–40
36. Salameh HAB, Krunz MM, Younis O (2009) MAC protocol for opportunistic cognitive radio networks with soft guarantees. IEEE Trans Mobile Comput 8(10):1339–1352
37. Su H, Zhang X (2007) The cognitive radio based multi-channel MAC protocols for wireless ad hoc networks. In: Proceedings of IEEE GLOBECOM, Washington DC, USA
38. Bukhari SHR, Siraj S, Rehmani MH (2016) PRACB: A novel channel bonding algorithm for cognitive radio sensor networks. IEEE Access 4:6950–6963
39. Foukalas F, Shakeri R, Khattab T (2018) Distributed power allocation for multi-flow carrier aggregation in heterogeneous cognitive cellular networks. IEEE Trans Wirel Commun 17(4):2486–2498
40. Hsu AC, Wei DSL, Kuo CCJ (2007) A cognitive MAC protocol using statistical channel allocation for wireless ad-hoc networks. In: Proceedings of IEEE WCNC, Hongkong, China
41. Shajaiah H, Abdelhadi A, Clancy C (2017) Resource allocation with carrier aggregation in cellular networks: optimality and spectrum sharing using C++ and MATLAB. Springer, Cham
42. Jiao L, Li FY, Pla V (2012) Modeling and performance analysis of channel assembling in multichannel cognitive radio networks with spectrum adaptation. IEEE Trans Veh Technol 61(6):2686–2697
43. Jiao L, Song E, Pla V, Li FY (2013) Capacity upper bound of channel assembling in cognitive radio networks with quasistationary primary user activities. IEEE Trans Veh Technol 62(4):1849–1855
44. Shaat M, Bader F (2010) Computationally efficient power allocation algorithm in multicarrier based cognitive radio networks: OFDM and FBMC systems. EURASIP J Adv Signal Process 2010(1):1–13
45. Re ED, Argenti F, Ronga LS, Bianchi T, Suffritti R (2008) Power allocation strategy for cognitive radio terminals. In: Proceedings of CogART, Aalborg, Denmark
46. Haddad M, Hayar AM, Debbah M (2007) Optimal power allocation for cognitive radio based on a virtual noise threshold. In: Proceedings of ACM CWNETS, Vancouver, Canada
47. Jia J, Zhang Q, Shen X (2008) HC-MAC: a hardware-constrained cognitive MAC for efficient spectrum management. IEEE J Sel Areas Commun 26(1):106–117
48. Zhang R, Liang YC, Cui S (2010) Dynamic resource allocation in cognitive radio networks: a convex optimization perspective. IEEE Signal Process Mag 27(3):102–114
49. Federal Communications Commission (2003) Notice of inquiry and notice of proposed rulemaking: in the matter of establishment of an interference temperature metric to quantify and manage interference and to expand available unlicensed operation in certain fixed, mobile and satellite frequency bands. ET Docket No 03-237, FCC-03-289

50. Federal Communications Commission (2007) Order: in the matter of establishment of an interference temperature metric to quantify and manage interference and to expand available unlicensed operation in certain fixed, mobile and satellite frequency bands. ET Docket No 03-237, FCC 07-78
51. Urgaonkar R, Neely MJ (2009) Opportunistic scheduling with reliability guarantees in cognitive radio networks. IEEE Trans Mobile Comput 8(6):766–777
52. Boyd S, Vandenberghe L (2004) Convex optimization. Cambridge University Press, Cambridge
53. Akyildiz IF, Lee WY, Chowdhury K (2009) CRAHNs: cognitive radio ad hoc networks. Ad Hoc Netw 7(5):810–836
54. Martinez-Bauset J, Pla V, Pacheco-Paramo D (2009) Comments on "analysis of cognitive radio spectrum access with optimal channel reservation". IEEE Commun Lett 13(10):739
55. Wong EWM, Foh CH (2009) Analysis of cognitive radio spectrum access with finite user population. IEEE Commun Lett 13(5):294–296
56. Jiao L, Pla V, Li FY (2010) Analysis on channel bonding/aggregation for multi-channel cognitive radio network. In: Proceedings of European wireless, Lucca, Italy
57. Lee J, So J (2010) Analysis of cognitive radio networks with channel aggregation. In: Proceedings of IEEE WCNC, Sydney, Australia
58. Jiao L, Li FY, Pla V (2011) Greedy versus dynamic channel aggregation strategy in CRNs: Markov models and performance evaluation. LNCS 6827:22–31
59. Jiao L, Li FY, Pla V (2011) Dynamic channel aggregation strategies in cognitive radio networks with spectrum adaptation. In: Proceedings of IEEE GLOBECOM, Houston, Texas, USA
60. Ai S, Jiao L, Li FY, Radin M (2016) Channel aggregation with guard-band in D-OFDM based CRNs: Modeling and performance evaluation. In: Proceedings of IEEE WCNC, pp 1–6

Chapter 2
Markov Chain and Stationary Distribution

MC has been a valuable tool for analyzing the performance of complex stochastic systems since it was introduced by the Russian mathematician A. A. Markov (1856–1922) in the early 1900s. More and more system analyses have been carried out by using MC, including the analysis on CA and CF. In this chapter, we will briefly review the essential ingredients of MC that are necessary for the performance analysis presented in this book. A more comprehensive introduction of MC and its applications can be found in [1].

2.1 Basic Concept of MC

An MC is a model of stochastic process. A stochastic process can describe the behavior of a system by defining:

1. all states that the system may stay, and
2. all transitions that the system may evolve from one state to another over time.

Formally, a stochastic process can be described by a collection of random variables, $\{X_t, \ t \in T\}$, which are defined in a given probability space and indexed by the parameter t, to represent the evolution of a certain system with random values *over time*. T is defined as the time index set. The values of the random variable X_t are called states, and the set of all possible states forms the *state space* of the process, which is denoted by $\mathcal{S}$.

One of the simplest stochastic processes is the Markov process, where the future of the process does not depend on its past, but only on its present. More precisely, a stochastic process $\{X_t\}$ is a Markov process if for all integers n and for any sequence $t_0, t_1, \ldots, t_n, t_{n+1}$, with $t_0 \leq t_1 \leq \ldots \leq t_n \leq t_{n+1}$, we have $P(X_{t_{n+1}} = j | X_{t_n} = i_{t_n}, \ldots, X_{t_0} = i_{t_0}) = P(X_{t_{n+1}} = j | X_{t_n} = i_{t_n})$, where $j \in \mathcal{S}$ and $i_{t_0}, \ldots, i_{t_n} \in \mathcal{S}$.

© The Author(s), under exclusive license to Springer Nature Switzerland AG 2020
L. Jiao, *Channel Aggregation and Fragmentation for Traffic Flows*,
SpringerBriefs in Electrical and Computer Engineering,
https://doi.org/10.1007/978-3-030-33080-4_2

This property is usually called Markov property, or memoryless property. A Markov process $\{X_t\}$ is time homogeneous if its transitions are independent of time t, i.e., $P(X_t = j | X_{t_n} = i_n) = P(X_{(t-t_n)} = j | X_0 = i_n)$.

In a stochastic process, if the state space is discrete, the process is called a chain. An MC is defined as a Markov process that has a discrete (finite or countable) state space. In this chapter, without loss of generality, we suppose that the state space of the chain, $\mathcal{S}$, is a set of non-negative integers. Furthermore, if the time set T is discrete in MC, the process is a *discrete* time MC (DTMC), otherwise the MC is a *continuous* time MC (CTMC).

2.2 Introduction to DTMC

A DTMC has a discrete state space and is observed at a discrete set of time instants. Since time is slotted, the time set becomes $T = \{0, 1, \ldots\}$. Therefore, the random variable X_{t_n} can be indexed simply by X_n.

Definition 2.1 The Markov process $\{X_n\}, n \in \{0, 1, \ldots\}$ is a DTMC if for all time index n, we have $P(X_{n+1} = j | X_n = i_n, \ldots, X_0 = i_0) = P(X_{n+1} = j | X_n = i_n)$.

The transition probability, i.e., the conditional probability $P(X_{n+1} = j | X_n = i_n)$, is usually called single-step transition probability. If the process is homogeneous, we can simplify the single-step transition as:

$$p_{ij} = P(X_{n+1} = j | X_n = i) = P(X_1 = j | X_0 = i). \tag{2.1}$$

The matrix $\boldsymbol{P}$, composed by placing p_{ij} in row i and column j, is defined as the transition probability matrix. Note that the entries in $\boldsymbol{P}$ have the property:

$$\sum_j p_{ij} = 1,\ 0 \le p_{ij} \le 1,\ \forall\, i,\ j \in \mathcal{S}. \tag{2.2}$$

Similar to the single-step transition defined by Eq. (2.1), we define the k-step transition probability for homogeneous process as:

$$p_{ij}^k = P(X_{n+k} = j | X_n = i) = P(X_k = j | X_0 = i). \tag{2.3}$$

The whole set of p_{ij}^k constitute the k-step transition probability matrix $\boldsymbol{P}^{(k)}$, which can be calculated from the single-step transition probability matrix $\boldsymbol{P}$ by using the Chapman–Kolmogorov equation as shown in Eq. (2.4).

Definition 2.2 In a DTMC whose single-step transition probability matrix is $\boldsymbol{P}$, the k-step transition probability matrix $\boldsymbol{P}^{(k)}$ can be calculated as:

$$\boldsymbol{P}^{(k)} = \boldsymbol{P}^{(n)} \boldsymbol{P}^{(k-n)},\ \forall\, 0 < n < k, \tag{2.4}$$

or in a scalar notation form:

$$p_{ij}^k = \sum_{\forall m \in S} p_{im}^n p_{mj}^{k-n}, \ \forall \, 0 < n < k. \tag{2.5}$$

Indeed, in DTMCs we are often interested in calculating the probability of a state that the chain is in after a certain number of transitions from a given initial state. This can be readily obtained by applying the Chapman–Kolmogorov equation. Specifically, if we define $\pi_j(n)$ as the probability that an MC is in *State j* at time step n, then the probability of any one of the states that the MC can be in at time step n, can be defined by the vector $\pi(n) = [\pi_0(n), \ \pi_1(n), \ \ldots \pi_j(n), \ \ldots]$. Note that $\sum_{j \in S} \pi_j(n) = 1$. By applying Eq. (2.4), we have:

$$\pi(n) = \pi(n-1)P = \pi(n-2)PP = \ldots = \pi(0)P^{(n)}.$$

Based on the transition features of DTMC, we have the following definitions:

- *State j* is *accessible* from *State i* if $p_{ij}^k > 0, \forall k \in \{1, \ 2, \ \ldots\}$.
- Two states *communicate* if they can access each other.
- If two states communicate, they are in the same *class*.
- An MC is said to be *irreducible* if it has only one class.

Definition 2.3 In a DTMC whose transition probability matrix is P, the vector $\pi = [\pi_0, \ \pi_1, \ \ldots, \ \pi_j, \ \ldots]$ is the stationary distribution if and only if:

$$\pi = \pi P, \ 0 \le \pi_j \le 1, \ \sum_j \pi_j = 1, \ \forall j \in S. \tag{2.6}$$

The stationary distribution of an MC is a distribution that stays unchanged as time progresses. Clearly, if the stationary distribution π is chosen to be the initial state distribution, i.e., $\pi(0) = \pi$, then after any possible number of transitions, the state distribution stays the same, i.e., $\pi(n) = \pi, \ \forall n \in \{1, \ 2, \ \ldots\}$.

We use a simple example to show a concrete instance of DTMC, and how its stationary probability distribution can be derived.

Example 2.1 It is shown in Fig. 2.1 a simple DTMC, and its transition matrix can be formulated as:

$$\mathbf{P} = \begin{bmatrix} p_{00} & p_{01} & p_{02} & p_{03} \\ p_{10} & p_{11} & p_{12} & p_{13} \\ p_{20} & p_{21} & p_{22} & p_{23} \\ p_{30} & p_{31} & p_{32} & p_{33} \end{bmatrix} = \begin{bmatrix} 3/4 & 1/4 & 0 & 0 \\ 4/5 & 0 & 1/5 & 0 \\ 0 & 1/3 & 1/3 & 1/3 \\ 0 & 0 & 1 & 0 \end{bmatrix}.$$

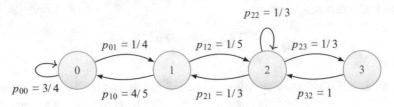

Fig. 2.1 A simple example of DTMC

If we denote the stationary probabilities of this DTMC by π_0, π_1, π_2, and π_3, then according to Definition 2.3, we have:

$$\begin{bmatrix} \pi_0 & \pi_1 & \pi_2 & \pi_3 \end{bmatrix} \begin{bmatrix} 3/4 & 1/4 & 0 & 0 \\ 4/5 & 0 & 1/5 & 0 \\ 0 & 1/3 & 1/3 & 1/3 \\ 0 & 0 & 1 & 0 \end{bmatrix} = \begin{bmatrix} \pi_0 & \pi_1 & \pi_2 & \pi_3 \end{bmatrix}. \tag{2.7}$$

Note that the state probabilities must sum up to 1, thus we have the following normalization equation:

$$\pi_0 + \pi_1 + \pi_2 + \pi_3 = 1. \tag{2.8}$$

Now we can calculate the stationary probability of this DTMC by solving Eqs. (2.7) and (2.8), and obtain:

$$\boldsymbol{\pi} = \begin{bmatrix} \pi_0 & \pi_1 & \pi_2 & \pi_3 \end{bmatrix} = \begin{bmatrix} 0.6400 & 0.2000 & 0.1200 & 0.0400 \end{bmatrix}.$$

Equation (2.7) can be written in a simplified way, which is called the balance equations of the MC, as follows:

$$1/4\pi_0 = 4/5\pi_1,$$
$$(4/5 + 1/5)\pi_1 = 1/4\pi_0 + 1/3\pi_2,$$
$$(1/3 + 1/3)\pi_2 = 1/5\pi_1 + \pi_3,$$
$$1/3\pi_2 = \pi_3. \tag{2.9}$$

The concept of Eq. (2.9) is that when a system becomes stable, all the probabilities of going out from a certain state, sum up to equal the sum of the probabilities of going into the same state. The balance equations can be easily established by observing the MC. For example, when the system is stable, the probability of going out from *State* 0, i.e., $1/4\pi_0$, should equal to the probability of going into *State* 0, i.e., $4/5\pi_1$. The same concept applies to all other states. Note that the self-loops are not included in the balance equations. This is because they will be canceled out anyways, as a loop goes both out from and back into the same state. Equation (2.9) can also be obtained by expanding Eq. (2.7) according to the rules of matrix multiplication. Indeed, Eqs. (2.9) and (2.7) are equivalent mathematically.

2.3 Introduction to CTMC

A DTMC has a discrete time space within which transitions occur. In other words, the system remains in a state for exactly one unit of time before transferring to the next state. On the contrary, in a CTMC, the system is allowed to spend a *continuous amount of time* being in a state, and thus has a continuous time space. Specifically speaking, suppose a CTMC enters into *State i*, where $i \in S$, then the length of time it spends in *State i* before it moves to the next state is a continuous positive random variable, which is called *sojourn time*. When the sojourn time terminates, the system makes a transition into the next state, say, *State j*, $j \in S$, with probability p_{ij}. Note that p_{ij} is different from the transition probability in the CTMC, as it only represents the probability that the system moves from *State i* to *State j* at the end of the sojourn time. Also note that $p_{ii} = 0$, as when the system moves from a state to itself, it only prolongs its sojourn time. Indeed, in CTMC, there is no self-transition.

Definition 2.4 A stochastic process $\{X_t, \ t \in T, \ t \geq 0\}$ with a discrete state space S is a CTMC, if for all $h \geq 0, s \geq 0, 0 \leq v < s$, and $\forall \, i, \ j \in S$,

$$P(X_{s+h} = j | X_s = i, \{X_v\}) = P(X_{s+h} = j | X_s = i) = P_{ij}(h), \tag{2.10}$$

where $\{X_v\}$ represents the set of past states before time s.

The definition of CTMC means that given the present state X_s, the future state X_{s+h} is independent of all the past states $\{X_v\}$. $P_{ij}(h)$ is the transition probability with which the chain transfers from *State i* to *State j*, within a duration of time h. If the conditional probability $P(X_{s+h} = j | X_s = i)$ depends only on the time interval h and is invariant for all s, we say that the CTMC is homogeneous [2]. We study only homogeneous CTMC in this book.

The transition probabilities in a CTMC are defined to be conditioned only on the current state of the process. This does not only mean that the CTMC's transition probabilities are independent from all the past states, but also means that they do not depend on the duration of time for which the process has been staying in the current state. The latter can be achieved by constraining the distribution of the sojourn time[1] so that the probability of the process staying in one state for a certain time period does not rely on how long it has been staying in that state. The exponential

---

[1]Without specifying the distribution of the sojourn time, the process becomes Semi-Markov process [1], and its future evolution depends on the current state of the process *and* on the length of the time for which the process has been in that state.

distribution has such a memoryless property. In more details, using S to denote the random variable of the sojourn time in a state, we have:

$$P(S > t_0 + t_1 | S > t_0) = \frac{P(S > t_0 + t_1, S > t_0)}{P(S > t_0)} = \frac{P(S > t_0 + t_1)}{P(S > t_0)}$$

$$= \frac{\int_{t_0+t_1}^{\infty} \lambda e^{-\lambda t} dt}{\int_{t_0}^{\infty} \lambda e^{-\lambda t} dt} = \frac{e^{-\lambda(t_0+t_1)}}{e^{-\lambda t_0}} = P(S > t_1). \qquad (2.11)$$

In Eq. (2.11), $P(S > t_1)$ is the probability of the system being in the current state for at least t_1, and $P(S > t_0 + t_1 | S > t_0)$ represents the probability that the system will be staying in the current state for at least another t_1, given that the system has already been in the state for at least t_0. When the sojourn time follows the exponential distribution, we have $P(S > t_0 + t_1 | S > t_0) = P(S > t_1)$, which means the condition that the system has already been staying in the current state for at least t_0 does not influence the probability of the system being in the same state for at least t_1. In other words, the sojourn time possesses memoryless property.

In CTMC, we use the concept of transition rate to describe the transitions of the process. The transition rates are essential elements in an infinitesimal generator matrix, denoted by Q, which can be derived from the transition probabilities of the CTMC.

If we denote the transition probability matrix by $P(h)$, which consists of $P_{ij}(h)$ defined in Eq. (2.10), then $P(0) = I$, where I is the identity matrix. This is because within *zero* amount of time, the process stays at the same state with probability 1. Suppose the sojourn time, S, follows an exponential distribution with rate λ_i, then the transition probability from *State i* to *State j* $(j \neq i)$ within time interval h, can be formulated as:

$$P_{ij}(h) = \int_0^h \lambda_{ij} e^{-\lambda_{ij}x} dx = 1 - e^{-\lambda_{ij}h}, \qquad (2.12)$$

where $\lambda_{ij} = \lambda_i p_{ij}$. Recall that p_{ij} is the probability that the system moves from *State i* to *State j* *at the end* of the sojourn time. If we denote the state space of all the destination states starting from *State i* by S_D, then $\sum_{j \in S_D} p_{ij} = 1$, and thus $\sum_{j \in S_D} \lambda_i p_{ij} = \lambda_i$. Also note that the integral in Eq. (2.12) is made from 0 to h, instead of from h to ∞. This is because the transition will always happen within the time interval h, indicating that the sojourn time $S \leq h$.

Now, we can define $Q = \lim_{h \to 0} \frac{P(h) - P(0)}{h}$, which is a matrix defined in infinitesimal time scale that describes the transition rate of the CTMC. Here we briefly summarize the derivation of the elements in Q.

Let q_{ij} be the elements in Q. We firstly consider the elements that are not located at the main diagonal of Q, i.e., $q_{ij}, \forall i \neq j$. Indeed $P_{ij}(0) = 0$ holds for all $i \neq j$, because given no time shift, the probability of transferring from *State i* to another

State j is zero. Therefore, q_{ij} can be derived as:

$$
\begin{aligned}
q_{ij} &= \lim_{h \to 0} \frac{P_{ij}(h) - P_{ij}(0)}{h} = \lim_{h \to 0} \frac{1 - e^{-\lambda_{ij}h} - 0}{h} \\
&= \lim_{h \to 0} \frac{1 - (1 - \lambda_{ij}h + O((-\lambda_{ij}h)^2))}{h} \\
&= \frac{\lambda_{ij}h}{h} = \lambda_{ij}, \ \forall i \neq j,
\end{aligned}
\tag{2.13}
$$

where we have applied Taylor series at 0 for $e^{-\lambda_{ij}h}$.

We now consider the elements of Q locating at the diagonal, i.e., q_{ii}. By definition,

$$
q_{ii} = \lim_{h \to 0} \frac{P_{ii}(h) - P_{ii}(0)}{h}.
\tag{2.14}
$$

As $P_{ii}(h) = 1 - \sum_{j \neq i} P_{ij}(h)$, and $P_{ii}(0) = 1$, Eq. (2.14) can be further calculated as:

$$
\begin{aligned}
q_{ii} &= \lim_{h \to 0} \frac{P_{ii}(h) - P_{ii}(0)}{h} = \lim_{h \to 0} \frac{1 - \sum_{j \neq i} P_{ij}(h) - 1}{h} \\
&= \lim_{h \to 0} \frac{-\sum_{j \neq i}(1 - e^{-\lambda_{ij}h})}{h} = \lim_{h \to 0} \frac{-\sum_{j \neq i} \lambda_{ij}h + \sum_{j \neq i} O((-\lambda_{ij}h)^2)}{h} \\
&= -\sum_{j \neq i} \lambda_{ij}.
\end{aligned}
\tag{2.15}
$$

Again, in the fourth step of Eq. (2.15), Taylor series is applied at 0 for $e^{-\lambda_{ij}h}$.

Given Eqs. (2.13) and (2.15), we see that $q_{ii} = -\sum_{j \neq i} q_{ij}$. Therefore, $\sum_j q_{ij} = 0$ holds for all i.

Similar to the DTMC, the Chapman–Kolmogorov equations can be utilized to calculate the transition probability.

Definition 2.5 In a CTMC, for all times s and t, the transition probability functions $P_{ij}(t + s)$ can be obtained from $P_{ik}(t)$ and $P_{kj}(s)$ as:

$$
P_{ij}(t + s) = \sum_k P_{ik}(t) P_{kj}(s),
\tag{2.16}
$$

where $P_{ik}(t)$ is the probability of transiting from *State* i to *State* k in time t and $P_{kj}(s)$ is the probability of transiting from *State* k to *State* j in the remaining time s.

Chapman–Kolmogorov equations have derivative forms, and we consider the forward equation as an example. By replacing s with a sufficiently small value, h, Eq. (2.16) becomes:

$$P_{ij}(t+h) = \sum_k P_{ik}(t)P_{kj}(h) = P_{ij}(t)P_{jj}(h) + \sum_{k \neq j} P_{ik}(t)P_{kj}(h). \qquad (2.17)$$

Expanding P_{jj} and P_{kj} by Taylor series, $P_{jj}(h) = 1 + q_{jj}h + O(h^2)$ and $P_{kj}(h) = q_{kj}h - O(h^2)$. Therefore, Eq. (2.17) can be further written as:

$$P_{ij}(t+h) = P_{ij}(t)P_{jj}(h) + \sum_{k \neq j} P_{ik}(t)P_{kj}(h)$$

$$= P_{ij}(t)(1 + q_{jj}h + O(h^2)) + \sum_{k \neq j} P_{ik}(t)(q_{kj}h - O(h^2)). \qquad (2.18)$$

Subtracting $P_{ij}(t)$ from both sides of Eq. (2.18), dividing the equation by h, and taking limit on both sides as $h \to 0$, we obtain:

$$\lim_{h \to 0} \frac{P_{ij}(t+h) - P_{ij}(t)}{h} = P_{ij}(t)q_{jj} + \sum_{k \neq j} P_{ik}(t)q_{kj} = \sum_k P_{ik}(t)q_{kj}. \qquad (2.19)$$

Equation (2.19) is exactly the Chapman–Kolmogorov forward equations.[2] Note that the left-hand side of Eq. (2.19) is the derivative of $P_{ij}(t)$. Writing the equation in the matrix form, we have:

$$\frac{dP(t)}{dt} = P(t)Q. \qquad (2.20)$$

Equation (2.20) implies that when a CTMC reaches statistical equilibrium, its transition probability $P(t)$ does not change over time. In this circumstance, $\frac{dP(t)}{dt}$ becomes 0 and thus each row in $P(t)$ becomes the stationary distribution.

Definition 2.6 In a CTMC with transition matrix of rates being Q, the vector $\pi = [\pi_0, \pi_1, \ldots \pi_j, \ldots]$ is considered to be the stationary distribution if and only if:

$$0 = \pi Q, \ 0 \leq \pi_j \leq 1, \ \sum_j \pi_j = 1, \ \forall j \in \mathcal{S}. \qquad (2.21)$$

We study the process of deriving the stationary probability distribution of a CTMC by examining the following simple example.

---

[2]Here we consider the time order as $0 \to t \to h$ and thus the forward equations apply. Similarly, when we consider $0 \to h \to t$, we can have the backward equations.

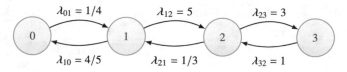

Fig. 2.2 A simple example of CTMC

Example 2.2 Figure 2.2 illustrates a simple CTMC. Given the transition rates λ_{ij}, the transition matrix Q can be formulated as:

$$Q = \begin{bmatrix} q_{00} & q_{01} & q_{02} & q_{03} \\ q_{10} & q_{11} & q_{12} & q_{13} \\ q_{20} & q_{21} & q_{22} & q_{23} \\ q_{30} & q_{31} & q_{32} & q_{33} \end{bmatrix} = \begin{bmatrix} -1/4 & 1/4 & 0 & 0 \\ 4/5 & -5.8 & 5 & 0 \\ 0 & 1/3 & -10/3 & 3 \\ 0 & 0 & 1 & -1 \end{bmatrix}.$$

Denoting the stationary probabilities of this CTMC as π_0, π_1, π_2,, and π_3, then by Definition 2.6, we have:

$$\begin{bmatrix} \pi_0 & \pi_1 & \pi_2 & \pi_3 \end{bmatrix} \begin{bmatrix} -1/4 & 1/4 & 0 & 0 \\ 4/5 & 5.8 & 5 & 0 \\ 0 & 1/3 & -10/3 & 3 \\ 0 & 0 & 1 & -1 \end{bmatrix} = \begin{bmatrix} 0 & 0 & 0 & 0 \end{bmatrix}. \qquad (2.22)$$

Expanding Eq. (2.22) by the matrix multiplication rule, we obtain:

$$4/5\pi_1 - 1/4\pi_0 = 0,$$
$$1/4\pi_0 - 5.8\pi_1 + 1/3\pi_2 = 0,$$
$$5\pi_1 - 10/3\pi_2 + \pi_3 = 0,$$
$$3\pi_2 - \pi_3 = 0. \qquad (2.23)$$

Besides, the state probabilities follow the normalization equation, i.e.,

$$\pi_0 + \pi_1 + \pi_2 + \pi_3 = 1. \qquad (2.24)$$

By solving Eqs. (2.23) and (2.24), we obtain the stationary probability of this CTMC as:

$$\pi = \begin{bmatrix} \pi_0 & \pi_1 & \pi_2 & \pi_3 \end{bmatrix} = \begin{bmatrix} 0.0498 & 0.0156 & 0.2336 & 0.7009 \end{bmatrix}.$$

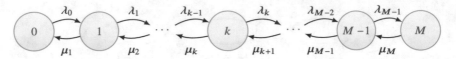

Fig. 2.3 A general birth and death process

Similar to the DTMC case, we can construct the balance equations for the CTMC as:

$$1/4\pi_0 = 4/5\pi_1,$$

$$(4/5 + 5)\pi_1 = 1/4\pi_0 + 1/3\pi_2,$$

$$(1/3 + 3)\pi_2 = 5\pi_1 + \pi_3,$$

$$3\pi_2 = \pi_3. \tag{2.25}$$

The balance equations are equivalent to Eq. (2.23), and can be used to calculate the stationary probabilities for the CTMC.

Example 2.3 Figure 2.3 shows an MC for a birth and death process (BDP). In the BDP, the states represent the number of items in a population. The state transitions have two types, namely "births" and "deaths". When birth happens, the number increases by one. On the contrary, when "death" happens, the number decreases by one. The transition rates are indicated in the figure.

The stationary state probabilities can be obtained by solving the following balance equations,

$$\lambda_0\pi_0 = \mu_1\pi_1,$$

$$(\lambda_k + \mu_k)\pi_k = \lambda_{k-1}\pi_{k-1} + \mu_{k+1}\pi_{k+1}, \quad 0 < k < M,$$

$$\lambda_{M-1}\pi_{M-1} = \mu_M\pi_M, \tag{2.26}$$

and the normalization equation, i.e., $\sum_j \pi_j = 1$, $j \in \{0, 1, \ldots, M\}$. Equation (2.26) can be simplified by canceling out the equal components on both sides of the equations. Take the balance equation for *State* 1 as an example, as $\lambda_0\pi_0 = \mu_1\pi_1$, the equation $(\lambda_1 + \mu_1)\pi_1 = \lambda_0\pi_0 + \mu_2\pi_2$ can be simplified to be $\lambda_1\pi_1 = \mu_2\pi_2$, hence π_2 can be written as a function of π_1, which, in turn, can be written as a function of π_0. Indeed, by simplifying Eq. (2.26), we can write each of the stationary probabilities, π_k, $0 < k \leq M$, as a function of π_0, as shown in Eq. (2.27):

$$\lambda_0 \pi_0 = \mu_1 \pi_1 \implies \pi_1 = \frac{\lambda_0}{\mu_1} \pi_0,$$

$$\lambda_1 \pi_1 = \mu_2 \pi_2 \implies \pi_2 = \frac{\lambda_1}{\mu_2} \pi_1 = \frac{\lambda_0 \lambda_1}{\mu_1 \mu_2} \pi_0,$$

$$\lambda_2 \pi_2 = \mu_3 \pi_3 \implies \pi_3 = \frac{\lambda_2}{\mu_3} \pi_2 = \frac{\lambda_0 \lambda_1 \lambda_2}{\mu_1 \mu_2 \mu_3} \pi_0,$$

$$\cdots,$$

$$\lambda_{k-1} \pi_{k-1} = \mu_k \pi_k \implies \pi_k = \frac{\lambda_{k-1}}{\mu_k} \pi_{k-1} = \frac{\prod_{i=0}^{k-1} \lambda_i}{\prod_{i=1}^{k} \mu_i} \pi_0,$$

$$\cdots,$$

$$\lambda_{M-1} \pi_{M-1} = \mu_M \pi_M \implies \pi_M = \frac{\lambda_{M-1}}{\mu_M} \pi_{M-1} = \frac{\prod_{i=0}^{M-1} \lambda_i}{\prod_{i=1}^{M} \mu_i} \pi_0. \tag{2.27}$$

Based on Eq. (2.27), the sum of the state probabilities is:

$$\sum_{k=0}^{M} \pi_k = \pi_0 \left(1 + \frac{\lambda_0}{\mu_1} + \ldots + \frac{\prod_{i=0}^{M-1} \lambda_i}{\prod_{i=1}^{M} \mu_i} \right) = \pi_0 \left(1 + \sum_{j=1}^{M} \frac{\prod_{i=0}^{j-1} \lambda_i}{\prod_{i=1}^{j} \mu_i} \right). \tag{2.28}$$

As $\sum_{k=0}^{M} \pi_k = 1$, we can calculate the stationary probabilities to be:

$$\pi_0 = \left(1 + \sum_{j=1}^{M} \frac{\prod_{i=0}^{j-1} \lambda_i}{\prod_{i=1}^{j} \mu_i} \right)^{-1}, \tag{2.29}$$

$$\pi_k = \frac{\prod_{i=0}^{k-1} \lambda_i}{\prod_{i=1}^{k} \mu_i} \left(1 + \sum_{j=1}^{M} \frac{\prod_{i=0}^{j-1} \lambda_i}{\prod_{i=1}^{j} \mu_i} \right)^{-1}, \quad 0 < k \leq M. \tag{2.30}$$

The way to solve for the stationary probabilities of the BDP will be utilized extensively in the remaining chapters.

Remarks

In DTMC, the elements in the transition matrix are probabilities that take values from [0, 1]. However, in CTMC, the elements in the transition matrix are rates. The rates on the diagonal of the matrix can take any values from $(-\infty, 0]$, whereas the rates located on the non-diagonal positions take values from $[0, +\infty)$.

We *never* associate self-loops with states in a CTMC, as opposed to its discrete-time counterpart, DTMC.

There are several methods that can be used to calculate the stationary distributions for CTMC and DTMC. When the state space becomes large, the task of calculating stationary distribution can be challenging. In such cases, approximation and asymptotic methods can be applied to simplify the computing process [1, 3].

References

1. Nelson R (2013) Probability, stochastic processes, and queueing theory: the mathematics of computer performance modeling. Springer Science & Business Media, New York
2. Anderson WJ (2012) Continuous-time Markov chains: an applications-oriented approach. Springer Science & Business Media, New York
3. Peng B (2004) Convergence, rank reduction and bounds for the stationary analysis of Markov chains. PhD dissertation, North Carolina State University

Chapter 3
Markov Chain Analysis of CA and CF with a Single Type of Users

In this chapter, we study the impact of CA and CF on traffic flows in the simplest system where there is one single type of flows generated by one single type of users. We will use CTMC to model the system, and the goal is to deliver the elementary concept of CTMC analysis for a system with CA and CF.

In the following sections, we firstly explain different flow types, and then present the system configurations and assumptions for the modeling process. Thereafter, we elaborate on the system modeling process itself for different flow types.

3.1 Flow Types and System Configurations

3.1.1 Real-Time and Elastic Traffic Flows

We consider two types of traffic flows in this chapter. They are real-time flows and elastic flows. Real-time flows describe traffic types such as Internet telephony or video calls, and elastic flows represent file downloading or traditional web browsing. Different types of traffic flows have distinct characteristics when CA and CF are applied [1, 2].

For elastic traffic, such as file downloading flows, transmission data rate is one of the most important factors that affect the transmission time. Therefore, in a system with CA, flows that are able to employ multiple channels for transmission can gain higher transmission data rate and thus shorter transmission time. Correspondingly, a system with CF provides possibilities for flows to utilize only a portion of a channel, thus allows lower data rate that results in longer transmission time. If CF is adopted together with CA in a system, the advantage, in addition to the improved data rate, is that the granularity of channel utilization becomes finer and more flows can be accommodated by the system as long as the minimum QoS criteria is still

L. Jiao, *Channel Aggregation and Fragmentation for Traffic Flows*,
SpringerBriefs in Electrical and Computer Engineering,
https://doi.org/10.1007/978-3-030-33080-4_3

maintained for each flow. In short, the service rate of an elastic flow is proportional to the number of channels that a flow occupies.

For real-time flows, as long as the required QoS is fulfilled, providing higher data rate will not shorten the conversation time. Indeed, the time duration for a real-time conversation is determined by the end users' demand and intention, and the service rate of a flow is no longer proportional to the number of channels that the flow occupies.

In this book, we study mathematical models corresponding to specific flow types. To avoid any confusion, the flow types will be specified when a mathematical model is introduced.

3.1.2 System Configuration and Basic Assumptions

In Chap. 2, we introduced two types of MCs, namely the DTMC and the CTMC. The study in this book will adopt CTMC, as it is carried out at flow level which is normally described in continuous time. DTMC is mostly applied in modeling a time-slotted system, such as a time-slotted MAC protocol at packet level [3, 4].

The statistic features of a certain type of flow are usually described by probability functions. In this book, unless otherwise stated, we assume the arrival process of the traffic flows to follow a Poisson process,[1] and the service time of flows to follow an exponential distribution. The benefit is that the system can be considered as memoryless by these assumptions. However, when we study the system performance by applying measurement-based distributions, as in Sect. 4.2, we will relax the assumption to general distributions for traffic flows.

To support a system with CA and CF, we assume there is a certain physical layer and MAC layer scheme that can deploy CA and CF for traffic flows. We also ignore the technique details, such as the water-filling algorithm, adopted by resource allocation schemes.

We assume that channels are homogeneous and different channels have the same long-term statistic features for traffic flows. With this assumption, we only need to care about the total number of channels that a flow utilizes, without tracking the details of each specific channel. Understandably, if channels are heterogeneous, each individual channel has its own characteristics, and following these channel details will increase the complexity of the system model significantly, which does not help with understanding the basic concept of system modeling.

Besides the above assumptions, we suppose the system has only one type of flows. In other words, the real-time and elastic flows do not co-exist in the system. Specifically in this chapter, we study the simplest case where there is only one single type of flows generated by one single type of users (single-flow single-user). Admittedly, this scenario barely exists in real-life communication systems.

---

[1]In Poisson process, the *time interval between events* follows exponential distribution.

However, it is a perfect toy-example for readers to easily understand the modeling process using MC analysis. More complicated models, where there is one single type of flows generated by multiple types of users with different channel access priorities (single-flow multi-user), are to be studied in the next chapter.

3.2 Analytical Models Based on CTMC

In this section, we use CTMCs to model different CA and CF strategies in the single-flow single-user case. To set a benchmark for modeling, we first present the analysis for the traditional strategy that is without CA and CF.

3.2.1 System Model without CA and CF

In a system without CA and CF, each flow always occupies one channel, regardless of the flow type it belongs to. We analyze real-time flows first. Consider a communication system with M channels. Suppose the arrival of flows follows a Poisson process with parameter λ, and we assume that there are an infinite number of users generating traffic flows so that the parameter of the arrival process does not vary with the number of ongoing flows.[2] We also assume that the service time of a flow follows an exponential distribution with service rate μ. As each flow in the system occupies one and only one channel, the access strategy for the system is to always accommodate an arrived flow as long as there is at least one vacant channel. A CTMC as simple as illustrated in Fig. 3.1 can describe such a system. In general, the state of the CTMC is defined to be the number of flows in the system, and is denoted by i. The transition from *State i* to *State i* $+ 1$ is λ, $\forall i < M$, and the transition from *State i* to *State i* $- 1$ is $i\mu$, $\forall i > 0$. The service rate for *State i* is given by $i\mu$ because there are i independent flows in the system, each of which occupies one single channel and has its service rate being μ. Indeed, Fig. 3.1 shows a concrete example of such a CTMC with $M = 3$.

If we denote the transition rate matrix of the CTMC by Q, then each element of Q represents the transition rate from one state to another. If we further define π as the stationary probability vector of the CTMC, then each stationary state probability, $\pi(i)$, can be calculated by solving the equation group composed of the global balance equation, $0 = \pi Q$, and the normalization equation, $\sum_i \pi(i) = 1$.

We can now move on to calculate the system parameters, namely, the capacity ρ and the blocking probability p_b.

The capacity of the system, ρ, is defined as the average number of flow completions per time unit [1]. It is calculated by summing up the products of state

---

[2]The case where there are a finite number of users will be studied later in Sect. 3.2.6.

Fig. 3.1 The CTMC without
CA and CF when $M = 3$

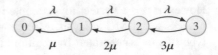

probabilities and their corresponding service rates, as shown in Eq. (3.1):

$$\rho = \sum_i i\mu\pi(i). \tag{3.1}$$

The blocking probability, p_b, refers to the probability that a flow arrival is blocked and lost due to insufficient channel resources. In this case, $p_b = \pi(M)$.

We now analyze for elastic flows. We keep the assumption that the service time of a flow transmitting on one single channel follows the exponential distribution with service rate μ. Though an elastic flow is able to obtain higher service rate if more channels are utilized for transmission, in the present system where CA and CF are not enabled, each flow is allowed to occupy only one channel. Therefore, when the CTMC is in *State i*, it means there are i ongoing flows with i channels being occupied, and the service rate of the state is $i\mu$, which is the same as that in the real-time flow case. Indeed, the CTMCs for elastic flows and real-time flows are identical in systems without CA and CF, since each flow is allowed to occupy only one channel at one time.

3.2.2   System Model with CA and CF

When CA and CF are enabled, the system model is quite different from the benchmark model introduced in Sect. 3.2.1. We study the modeling process for such systems in this section.

3.2.2.1   Models for Elastic Flows

We examine the elastic flows first. Similar to the benchmark model, for the single-flow single-user case, we consider a communication system with M channels. The arrival of the flows follows a Poisson process with parameter λ, and the service time of the elastic flow *on one channel* follows an exponential distribution with service rate μ. Let $N \in \mathbb{R}^+$ be the number of channels that a flow adopts, and let $W \in \mathbb{R}^+$ and $V \in \mathbb{R}^+$ represent the respective minimum and maximum numbers of aggregated channels allowed for the flow, then $0 < W \leq N \leq V \leq M$. If the channels in the communication system are homogeneous, the service rate for a flow with N channels is $N\mu$. Note that we have assumed the service rate for a flow to be linear with the number of its aggregated channels. Indeed, in reality, depending on the system configurations, this may not be the case. When the service rate is not linearly proportional to the number of occupied channels [5], it needs to be adjusted according to a function that maps the occupied channels to the actual service rate. However, in this study, we keep the linear assumption for simplicity.

In the following paragraphs, we will present three strategies and their CTMC models.

Strategy EFAFS(W, V) We assume that the users in this communication system can adjust the number of channels an ongoing flow occupies, and that the ongoing flows operate in the following manner:

- The ongoing flows always utilize as many channels as they are able to.
- The ongoing flows always equally share the channels.
- The ongoing flows always share their occupied channels with a new arrival by adjusting the number of their own channels.

Such a channel access strategy means, if one or more channels become idle due to a departure of an ongoing flow, the remaining ongoing flows will equally share the newly available channels, with the constraint that the number of channels each flow occupies is not greater than V. It also means upon an arrival of a flow, the new flow will be allowed to commence if and only if the number of channels per flow is not lower than W after channel sharing. One may have noticed that in order for a flow to equally share channel resources, the flow must be able to assemble a non-integer number of channels with the help from CF. Though in reality, due to the constraint in hardware, it might be impossible to achieve an absolutely equal channel sharing among all ongoing flows, we assume in this study that the granularity of channel sharing is fine enough to keep the statistics of flows. We name this channel access strategy as "extended full adaptation and full sharing strategy," and denote it as EFAFS(W, V).

We are to establish a CTMC to analyze the behavior of the system with EFAFS(W, V). Let i be the number of ongoing flows in the system and the index of the states in the CTMC. The transition of the CTMC from *State i* to *State i* + 1 is made upon a new flow arrival and the transition rate is λ if $i < \lfloor \frac{M}{W} \rfloor$. $\lfloor \frac{M}{W} \rfloor$ indicates the maximum number of ongoing flows that can be accommodated in the system,[3] hence $i \leq \lfloor \frac{M}{W} \rfloor$. Similarly, the transition of the CTMC from *State i* to *State i* − 1 is made when an ongoing flow finishes and leaves the system, and the transition rate is equal to the service rate for *State i*. More specifically, if the number of channels that all the ongoing flows occupy is less than the maximum number of channels in the communication system, i.e., $Vi < M$, then the service rate is $Vi\mu$. If all the

---

[3]In this book, we assume the maximum number of ongoing flows is bounded by $\lfloor \frac{M}{W} \rfloor$ for simplicity. In reality, the number may be greater or less than $\lfloor \frac{M}{W} \rfloor$. For example, due to the utilization of guard bands in between the original channels for transmission, the overall data rate of the system with CA and CF may be higher than that of the system without CA and CF. In such a case, a greater number of simultaneous flows can be supported as the increased overall data rate is able to support more flows. One can adjust the analytical model by extending or reducing the length of the CTMC to fit the system in reality.

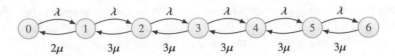

Fig. 3.2 The CTMC of EFAFS(0.5, 2) with $M = 3$ for elastic flows

M channels are occupied by the ongoing flows already, i.e., $Vi \geq M$, then due to the assumption that the maximum number of channels in the system cannot be extended, the overall service rate becomes $M\mu$.

Figure 3.2 illustrates such a CTMC with $W = 0.5$, $V = 2$, and $M = 3$. In this system, each flow requires at least 0.5 channel for communication. Therefore, the maximum number of ongoing flows is 6, resulting in a 7-state CTMC. Note that the service rate at *State* 1 is 2μ, because the system at this state has only one flow which occupies at most two channels. When $i > 1$, the service rate becomes 3μ, as all three channels can be fully utilized by flows.

Based on the state transitions presented above, we can compose the transition rate matrix Q of the CTMC, and calculate the stationary state probability, $\pi(i)$, for each of the states in the CTMC. Both Q and $\pi(i)$ are further used for calculating the key parameters of the communication system.

The capacity of the system, ρ_e, is defined as:

$$\rho_e = \sum_{i,\, iV \leq M} \pi(i) iV\mu + \sum_{i,\, iV > M} \pi(i) M\mu. \tag{3.2}$$

The first part on the right-hand side of Eq. (3.2) is the average service rate when the ongoing flows have not used up all channels in the system, whereas the second part represents the average service rate when all channels are occupied.

The blocking probability of the system is, $p_{b_e} = \pi\left(\lfloor \frac{M}{W} \rfloor\right)$.

Another system parameter of interest is the number of channels that each flow utilizes after CF and CA. We denote it by $\bar{N}_e$, and $\bar{N}_e$ is defined to be the average number of channels that all flows occupy divided by the average number of flows in the system, as shown in Eq. (3.3),

$$\bar{N}_e = \frac{E(No.\ of\ channels\ utilized)}{E(No.\ of\ flows)} = \frac{\sum_{i,\, 0 < iV \leq M} \pi(i) iV + \sum_{i,\, iV > M} \pi(i) M}{\sum_i \pi(i) i}. \tag{3.3}$$

Strategy Greedy(W, V) To illustrate the importance of allowing new arrivals to commence, we now model a strategy where ongoing flows will *not* share their occupied channels with the new arrival when the number of idle channels is insufficient for the newcomer to commence. We name this strategy Greedy(W, V). In Greedy(W, V), the maximum number of ongoing flows equals $\lfloor \frac{M}{V} \rfloor + 1$ if $M - V\lfloor \frac{M}{V} \rfloor \geq W$, and $\lfloor \frac{M}{V} \rfloor$ otherwise. The transitions of the CTMC are similar to those

Fig. 3.3 The CTMC of
Greedy(0.5, 2) with $M = 3$
for elastic flows

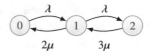

of EFAFS(W, V), and the length of the CTMC reduces due to the greedy behavior
of the ongoing flows. Figure 3.3 illustrates the CTMC for the Greedy(W, V) system
when $M = 3$, $W = 0.5$, and $V = 2$.

In Fig. 3.3, when there is only one flow in the system, i.e., when the CTMC is
at *State* 1, the service rate is 2μ as the flow can utilize at maximum two channels.
When a new flow arrives, the one idle channel allows the system to accommodate
the new flow as the requirement for the new flow to commence is that at least a half
of a channel is available. Upon the new flow's arrival, the system moves from *State* 1
to *State* 2. Because now the ongoing flows take up all channels, the service rate at
State 2 is 3μ. The completion of either of the two flows will result in the system
moving back to *State* 1. When the system reaches *State* 2, it will not accommodate
any new flows since all the channels are being used and the ongoing flows do not
share their own channels. The new flows will thus be blocked. The capacity of the
system and the indicator of the number of channels that a flow occupies can be
calculated in the same way as those in the EFAFS(W, V) system.

Strategy Dy(W, V) Both strategies EFAFS(W, V) and Greedy(W, V) deal with
flows that equally share channels. In the following so-called Dy(W, V) strategy, we
consider flows that do *not* share their channels equally, i.e., they can have different
numbers of channels.

As in the strategy EFAFS, we use $N \in \mathbb{R}^+$ to denote the number of channels that
a flow utilizes, and $W \in \mathbb{R}^+$ and $V \in \mathbb{R}^+$ to represent the minimum and maximum
numbers of channels a flow can occupy, respectively. We have $0 < W \leq N \leq V \leq$
M. Consider the arrival event first. Suppose a flow tries to access channels in the
Dy(W, V) system, the flow can commence directly if the number of idle channels,
denoted by I_c, is no less than the minimum number, W, at that moment. In this
case, the number of channels that the new flow can utilize is $\min\{V, I_c\}$. In the case
that there are not enough idle channels in the system for a newly arrived flow to
commence, ongoing flows will share their occupied channels with the new flow,
as long as each ongoing flow and the new flow can get at least W channels after
sharing. Channel sharing upon arrival in the Dy(W, V) system is not equal. Instead,
it obeys a rule that is based on priority. Specifically speaking, when a new flow needs
ongoing flows to share channels with it, and if there are more than one ongoing
flows existing in the system, the ongoing flow that occupies the most channels will
donate first. If the flow with the most channels cannot provide enough channels to
accommodate the new flow, the one with the second most channels will then share its
channels, and so on. If the number of idle channels plus the channels to be released
by ongoing flows surpasses the minimum number W, the newly arrived flow can
join the network successfully. Otherwise, the new arrival will be blocked.

The departure event in the Dy(W, V) system also involves different scenarios. A
general rule is that, when channel resource becomes available upon a flow departure,

the flow with the minimum number of channels will utilize the resource first, and so on. Table 3.1 lists the typical scenarios for both arrival and departure events in the first column.

To construct the CTMC for this strategy, we first define the states of the CTMC model as $x = (j_W, \ldots, j_k, \ldots, j_V)$, where j_k is the number of flows that utilize $k = W, 2W, \ldots,$ or V channels. Note that in this definition, for the sake of simplicity, we assumed N, M, and V to be integer multiples[4] of W. We then denote the total number of channels that are being used in the system at *State* x as $b(x) = \sum_{k=W}^{V} k j_k$, and the set of feasible states as $S = \{x \mid j_W, \ldots, j_V \geq 0; \ b(x) \leq M\}$. Then the transitions between states in this CTMC upon different conditions can be described by Columns 2, 3, and 4 in Table 3.1.

Because the Dy(W, V) system involves not only the number of ongoing flows but also the number of channels each flow occupies, the transition of its CTMC is more complicated than that for EFAFS(W, V) and Greedy(W, V). The complexity of Dy(W, V) also lies in the fact that the destination state of the system upon an arrival or a departure scenario may involve multiple different system conditions. Take the first departure scenario in Table 3.1 as an example. It involves the following three different conditions:

1. $j_k = 1, k < V; j_m = 0, \forall m < V$ and $m \neq k; V > W$. This describes the condition that in the system, there exists only one flow with $k < V$ channels, and at the time the flow departs, each of the remaining ongoing flows has no less than V channels.
2. $j_k > 0, k = V; j_m = 0, \forall m < V; V > W$. This condition describes that all flows in the system, including the ongoing ones and the departing one, each occupies V channels.
3. $j_k > 0, k = W = V$. This is a condition where all flows occupy a constant number of flows all the time.

Under any one of the above conditions, when the flow with k channels leaves the system, the remaining ongoing flows are not allowed to use the vacant channel released by the departed flow, as each of the remaining ongoing flows has already taken up V channels, reaching the maximum number of channels allowed to a flow. Upon the departure of the flow, the system will move to the destination state of $(j_W, \ldots, j_k - 1, \ldots, j_V)$.

Other transitions in Table 3.1 can be derived following the same concept.

To further understand the Dy(W, V) strategy, we study the CTMC with $W = 0.5$, $V = 2$, and $M = 3$, as plotted in Fig. 3.4. We take *State* $(1,1,1,0)$ as an example to explain the transitions. *State* $(1,1,1,0)$ means there are three flows in the system and they occupy 0.5, 1, and 1.5 channels, respectively. When a flow arrives, the ongoing flow that occupies 1.5 channels will donate 0.5 channel to the new flow, and the

---

[4]If we relax the constraint to non-integer multiples of W, then the dimension of the state space will be higher, though the same analysis concept can be utilized in that situation with an increased complexity.

Table 3.1 Transitions from a generic state $x = (j_W, \ldots, j_V)$ of $\mathrm{Dy}(W, V)$

Activity	Destination state	Trans. rate	Conditions
Arrival. Enough idle channels exist	$(j_W, \ldots, j_k+1, \ldots, j_V)$	λ_S	$k = \min\{M - b(x), V\} \geq W$
Arrival. The flow with the maximum number of channels, m, donates channel(s) to the new flow	$(j_W+1, \ldots, j_n+1, \ldots, j_m-1, \ldots, j_V)$	λ_S	$m = \max\{r \mid j_r > 0, 2W \leq r \leq V\}; n = m - [W - (M - b(x))], W \leq n < m; V > W$
Departure. Other flows, if exist, cannot use the vacant channel(s)	$(j_W, \ldots, j_k-1, \ldots, j_V)$	$k j_k \mu_S$	$j_k = 1, k < V; j_m = 0, \forall m < V$ and $m \neq k; V > W$. Or $j_k > 0, k = V; j_m = 0, \forall m < V; V > W$. Or $j_k > 0, k = W = V$
Departure. The flow with the minimum number of channels, h, uses all the vacant channel(s)	$(j_W, \ldots, j_h-1, \ldots, j_i+1, \ldots, j_V)$	$k j_k \mu_S$	$j_k > 1; h = \min\{r \mid j_r > 0, W \leq r \leq V - W\}; l = k + h \leq V; V > W$. Or $j_k = 1; h = \min\{r \mid j_r > 0, r \in \{W, \ldots, k-1, k+1, \ldots, V - W\}\}; l = k + h \leq V; V > W$
...	...	$k j_k \mu_S$	...
Departure. All other flows with fewer than V channels use the vacant channel(s) and achieve the maximum number, V	$(0, \ldots, 0, \ldots, j_V + q)$	$k j_k \mu_S$	$q = \sum_{m=W}^{V-1} j_m - 1; V > W; k \geq \sum_{m=W}^{V-W}(V - m)j_m - (V - k)$

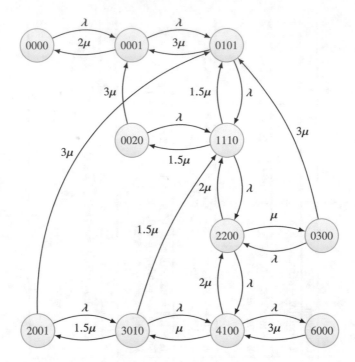

Fig. 3.4 The CTMC of Dy(0.5, 2) with $M = 3$ for elastic flows

result is that the donating flow occupies 1 channel after donating and the new flow is accommodated by the system with the donated 0.5 channel. As there are two flows in the system that occupies 0.5 channel, and two flows that are using 1 channel, the system has moved to the destination state of (2,2,0,0) upon the arrival of the new flow.

When a flow departs from (1,1,1,0), depending on which flow is departing, the destination state could be either (0,0,2,0) or (0,1,0,1). When the flow that occupies 1.5 channels departs, the released channels will be utilized by the flow with the least 0.5 channel, resulting in a flow that occupies two channels. In this case, the destination state is (0,1,0,1). Similarly, if the departing flow is the one with 0.5 channel or the one that occupies 1 channel, then after the released channel resource is reused by the remaining flows, the system ends up with two flows, each of which owns 1.5 channels. In this case, the system will move to the destination state of (0,0,2,0).

Analysis on the transitions between other states can be carried out similarly.

With the CTMC being constructed, and $\pi(x)$ being the stationary state probability of *State* x in the CTMC, the capacity of the Dy(W, V) system, ρ'_e, can be calculated by:

$$\rho'_e = \sum_{x \in \mathcal{S}} \sum_{k=W}^{V} k j_k \mu \pi(x). \tag{3.4}$$

The blocking probability of the system can be presented by:

$$p'_{b_e} = \sum_{\substack{x \in \mathcal{S}, \\ M-b(x)+\sum_{k=2W}^{V}(k-W)j_k < W}} \pi(x), \tag{3.5}$$

and the number of channels that a flow statistically utilizes, $\bar{N}'_e$, can be defined to be:

$$\bar{N}'_e = \frac{\sum_{x \in \mathcal{S}} \sum_{k=W}^{V} kj_k \pi(x)}{\sum_{x \in \mathcal{S}} \sum_{k=W}^{V} j_k \pi(x)}. \tag{3.6}$$

3.2.2.2 Models for Real-Time Flows

One of the key differences between real-time flows and elastic flows is that real-time flows maintain constant service rate. More precisely, even in a system with CF and CA, where more or less channels can be utilized by the traffic flow, the service rate of a real-time flow will not change as long as its minimum QoS requirement is satisfied. Due to the difference, the models for real-time flows are different from those for elastic flows. In this section, we establish real-time-flow models for the EFAFS(W, V) strategy and the Greedy(W, V) strategy.

In the EFAFS(W, V) strategy, ongoing flows share their own channels with a new arriving flow. The maximum number of flows is bounded by $\lfloor \frac{M}{W} \rfloor$. If the system is in *State i* with $i < \lfloor \frac{M}{W} \rfloor$, the transition of the CTMC from *State i* to *State i* + 1 can be made upon a new flow arriving and the transition rate is λ. Similarly, when an ongoing flow finishes and leaves the system, the transition of the CTMC is made from *State i* to *State i* − 1. As the service rate for each flow is μ, if there are N flows in the system, the service rate of the entire system is $N\mu$. Figure 3.5 illustrates a real-time-flow CTMC with $W = 0.5$, $V = 2$, and $M = 3$. The CTMC has in total 7 states, indicating that there can be at most six ongoing flows co-exist in the system. This is because there are in total 3 channels in the system and each flow requires at least 0.5 channel for communication. The service rate of the system is linearly proportional to the number of ongoing flows.

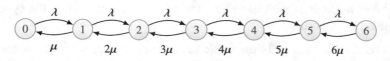

Fig. 3.5 The CTMC of EFAFS(0.5, 2) with $M = 3$ for real-time flows

Fig. 3.6 The CTMC of
Greedy(0.5, 2) with $M = 3$
for real-time flows

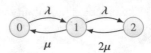

The capacity of the system, ρ_r, can be then defined as:

$$\rho_r = \sum_{i,\, i \leq M} \pi(i) i \mu. \tag{3.7}$$

The formulas for the blocking probability and the average number of channels that each flow utilizes are, in this case, the same as those for the elastic flows in the EFAFS(W, V) strategy.

For Greedy(W, V) with real-time flows, a CTMC with $M = 3$, $W = 0.5$, and $V = 2$ is illustrated in Fig. 3.6. We observe that the only difference between this CTMC and the one in Fig. 3.3 for elastic flows is the service rates. Therefore, if we replace the service rates with the new ones, we can calculate the system parameters for this real-time-flow CTMC in the same way as for the elastic traffic flows in the same strategy.

3.2.3   Numerical Results

In Sects. 3.2.1 and 3.2.2, we utilized CTMC to model six different systems, and they are:

1. The benchmark model for the system without CA and CF.
2. The elastic-flow model for the EFAFS(W, V) strategy.
3. The elastic-flow model for the Greedy(W, V) strategy.
4. The elastic-flow model for the Dy(W, V) strategy.
5. The real-time-flow model for the EFAFS(W, V) strategy.
6. The real-time-flow model for the Greedy(W, V) strategy.

Before we continue with further analyses, we display some numerical results from the mathematical models to further illustrate the behavior of the systems under various strategies for different types of traffic flows.

In order to obtain numerical results that can be fairly compared, the system is configured as follows:

- For both elastic and real-time traffic, in systems with or without CA and CF, and for all the three strategies, EFAFS(W, V), Greedy(W, V), and Dy(W, V): the maximum number of channels in the system is set to be $M = 3$.
- The service rate for a real-time flow is set to be 0.5 per flow, while the service rate for an elastic flow is set to be 0.5 per channel.

- In systems where CA and CF are enabled, we configure the minimum and maximum numbers of channels that one flow is allowed to occupy as $W = 0.5$ and $V = 2$.
- We plot the system capacity, the blocking probability, and the indicator for the number of channels per flow, as functions of the arrival rate, λ, which ranges from 0.2 to 0.9.
- The unit of the rate is the number of flows per time unit, and the time unit can be, e.g., in seconds or minutes, depending on the system to be modeled.

The results of elastic flows are plotted in Figs. 3.7, 3.8, and 3.9.

Figure 3.7 illustrates the system capacity of different strategies for elastic flows. All the capacities have an increasing trend with an increasing λ. The main reason is that when the system is far from saturation, more flows will be injected into

Fig. 3.7 The capacity for elastic flows with different CA and CF strategies

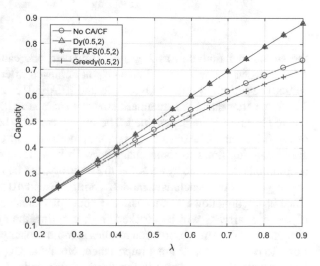

Fig. 3.8 The blocking probability for elastic flows with different CA and CF strategies

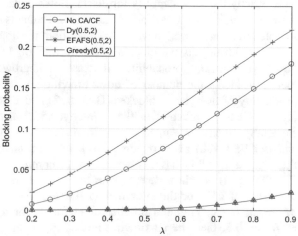

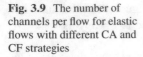

Fig. 3.9 The number of channels per flow for elastic flows with different CA and CF strategies

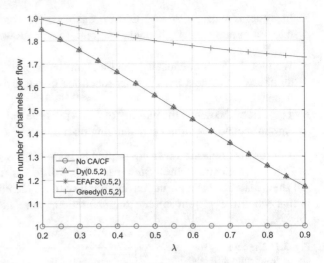

the system per time unit as λ grows. However, one can expect the trend to stop increasing when the saturated capacity is achieved after λ surpasses a particular value. The capacity values of Dy(0.5, 2) and EFAFS(0.5, 2) increase almost linearly, which is better than the benchmark strategy (named as "No CA/CF" in the figure) where CA and CF are not enabled. The improvement is benefited from the features of Dy(0.5, 2) and EFAFS(0.5, 2), i.e., they both support CA and CF, and they both allow ongoing flows to share channels with new flows so that new flows can be accommodated into the system. On the contrary, the capacity of Greedy(0.5, 2) is not as good as the benchmark model. Though Greedy(0.5, 2) supports CA and CF, since its ongoing flows do not share their occupied channels with new arrivals, most of the new arrivals will be blocked from entering the system, resulting in a low system capacity. This observation implies if more capacity is required, allowing new flows to commence is of great importance. Note that Dy(0.5, 2) and EFAFS(0.5, 2) have exactly the same capacity for elastic flows. Indeed, this is also the case with their blocking probabilities and the number of channels per flow. The reason is to be revealed in the next section where we talk about the upper bound of the capacity.

Figure 3.8 depicts the blocking probabilities of different strategies for elastic flows. The blocking probabilities increase as λ grows, because an increasing λ means an increasing number of flows arriving per unit time, and hence a higher probability of them being blocked. Dy(0.5, 2) and EFAFS(0.5, 2) have in general lower blocking probabilities than Greedy(0.5, 2) and the benchmark strategy, especially when λ is large. The main reasons are threefold. Firstly, Dy(0.5, 2) and EFAFS(0.5, 2) allow ongoing flows to share channels with newcomers, and therefore, a new flow has a better chance to commence. Secondly, Dy(0.5, 2) and EFAFS(0.5, 2) enable a flow to start with a half of a channel for transmission by virtue of CF. For the system with the maximum number of channels being $M = 3$, if the minimum number of channels for a flow to start transmission is $W = 0.5$, then the maximum number of ongoing flows in the system is 6, which is doubled compared with that in the benchmark strategy. In other words, Dy(0.5, 2) and EFAFS(0.5, 2) lower the threshold, W, for a flow to commence, and therefore, they are able to accommodate a larger number of flows, and thus achieve a higher system capacity. Finally, in Dy(0.5, 2) and EFAFS(0.5, 2), ongoing flows are allowed to keep aggregating channels released by departed flows, as long as its channel resources are within the maximum number V. For elastic flows, more channels means higher service rate. Therefore, flows can be finished faster, leaving more room to accommodate new arrivals.

The blocking probability of Greedy(0.5, 2) is greater than that of the benchmark strategy. This is due to its greedy nature that an ongoing flow can occupy multiple channels but will not share channels with the new arrival when the latter needs the shared channels to commence.

Figure 3.9 shows the number of channels per flow for elastic traffic. The trend in Dy(0.5, 2), EFAFS(0.5, 2), and Greedy(0.5, 2) is descending as λ increases. This is because more flows will be sharing the same limited channel resources when the arrival rate becomes larger. The number in the benchmark strategy is a constant though, because each flow always occupies one channel in this strategy. Among the strategies Dy(0.5, 2), EFAFS(0.5, 2), and Greedy(0.5, 2), the number in the Greedy strategy is the largest. This is because ongoing flows in this strategy do not share their channels with new arrivals, which allows the system to accommodate at most 2 flows simultaneously, and results in a generous resource occupancy for the ongoing flows.

Although within the range of λ that is shown in Fig. 3.9, one can observe the number of channels per flow in Dy(0.5, 2) and EFAFS(0.5, 2) being greater than that in the benchmark strategy. Indeed, when λ keeps on growing, these numbers will become closer to and eventually lower than that of the benchmark strategy. However, those cases are not of interest because the blocking probability will be too high for a reasonable communication system.[5] In general, Dy(0.5, 2) and EFAFS(0.5, 2) significantly outperform the benchmark strategy in terms of system performance for elastic flows.

Figures 3.10, 3.11, and 3.12 illustrate the system parameters for real-time traffic in different strategies.

In Fig. 3.10, the overall trend of the system capacity is ascending as the arrival rate λ increases. The capacity increases because more traffic is injected into the system. Again, EFAFS(0.5, 2) provides an improved system capacity in comparison with the benchmark strategy. The capacity of the Greedy(0.5, 2) strategy is the lowest, indicating that it is important for ongoing flows to share their channels with the new arrivals if an enhanced system capacity is to be expected.

Figure 3.11 shows that the real-time flows have a similar trend of blocking probability to the elastic traffic. The EFAFS(0.5, 2) has the lowest blocking

[5]We can observe in Fig. 3.8 that when $\lambda = 0.9$, the blocking probability of Greedy(0.5, 2) has already reached 22%, and the blocking probability of the benchmark strategy has surpassed 17% also. These values are indeed very high already.

Fig. 3.10 The capacity for
real-time flows with different
CA and CF strategies

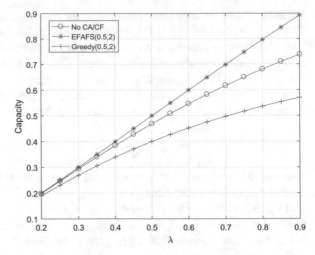

Fig. 3.11 The blocking
probability for real-time flows
with different CA and CF
strategies

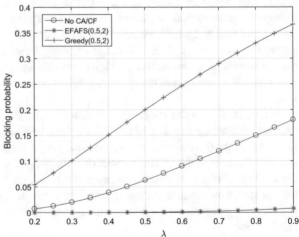

probability while the Greedy(0.5, 2) has the highest. Also notice that the blocking
probability of EFAFS(0.5, 2) for real-time flows is lower than that for elastic flows,
and the difference becomes more apparent when λ is larger. The main reason for
this is that these two cases have different service rates at the system level. The
service rate for elastic flows is determined by how many channels that all ongoing
flows jointly occupy. Therefore, the maximum service rate is 3μ, as demonstrated in
Fig. 3.2, which happens when all channels are being utilized by the ongoing flows.
For the real-time flows, the service rate is linearly proportional to the number of
ongoing flows in the system. Therefore, the maximum service rate is 6μ when the
number of flows accommodated in the system reaches the maximum, as shown in
Fig. 3.5. As most of the states in Fig. 3.2 for elastic flows have a lower service
rate than those in Fig. 3.5 for real-time flows, the blocking probability, i.e., the

Fig. 3.12 The number of channels per flow for real-time flows with different CA and CF strategies

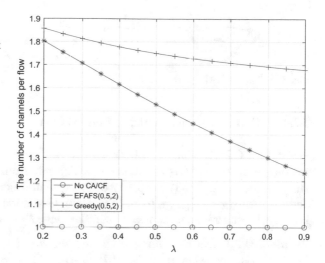

probability that the system is in the right-most state (e.g., *State* 6 in Fig. 3.2 and Fig. 3.5), is greater for elastic flows than for real-time flows.

From the number of channels per flow that is presented in Fig 3.12, we understand that both EFAFS(0.5, 2) and Greedy(0.5, 2) outperform the benchmark strategy within a reasonable range of blocking probability. Indeed, the service rate for real-time flows cannot be increased by CA and CF, but it does not stop a certain flow from utilizing more channels statistically. Statistically using more channels results in statistically increased data rates for real-time flows.

From the above illustrated numerical results, we observe that regardless of the traffic types, CA and CF can indeed improve the system performance for traffic flows, given the strategy being properly designed. Though the scenarios being studied so far are mere simple examples which can hardly be seen in real communication systems, the analytical process provides us with an initial comprehension of applying CTMC in modeling a certain channel access strategy. Besides, the promising results demonstrated by employing CA and CF are encouraging and motivate us to further explore the benefits that these techniques can bring to the system.

3.2.4 Capacity Upper Bound of Elastic Traffic Flows

As presented in the previous section, with various CA and CF strategies applied to traffic flows, a system can achieve capacities at different levels. Capacity improvement has been observed in the system with EFAFS(W, V) and the system with Dy(W, V). One question is, is there any capacity upper bound for the strategies with CA and CF being applied under the same system configurations? The answer is "Yes," and the upper bound is the one that can be achieved by EFAFS(W, V), with $V = M$ and W being set to be the minimum value satisfying the QoS requirement for the elastic flows. We will prove this statement in the following two subsections.

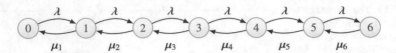

Fig. 3.13 The BDP with $K = 6$

3.2.4.1 Capacity Upper Bound of EFAFS(W, V)

We study a CTMC which is a BDP with $K + 1$ states. The transition rate from *State* $(k - 1)$ to *State* k is λ, and the transition rate from *State* k to *State* $k - 1$ is μ_k, $k \in \{1, \ldots, K\}$. The capacity of the system described by the BDP can be defined as $\rho_{BDP} = \sum_{k=1}^{K} \pi_k \mu_k$, with π_k being the stationary probability of *State* k. Figure 3.13 illustrates an example of the BDP with $K = 6$.

Proposition 3.1 *If a BDP has a fixed length and a fixed arrival rate, then* $\forall k \in \{1, \ldots, K\}$*, the capacity of the BDP,* ρ_{BDP}*, will increase monotonically as* μ_k *increases.*   □

Proof As the blocking probability of the system is π_K, $\rho_{BDP} = (1 - \pi_K)\lambda$. This is because the rate of flows being completed equals to the rate of arrival deducting the rate being blocked. As λ is a constant by definition, to prove that ρ_{BDP} increases with an increasing μ_k, we can prove that π_K decreases as μ_k increases.

In Chap. 2, we have introduced that the stationary probability of a CTMC can be calculated by solving the balance equations and the normalization equation of the CTMC. Following the same method, we can calculate the stationary distribution of the above mentioned BDP, and represent the stationary distribution of *State* k as a function of λ, μ_k, and the stationary probability of *State* 0, i.e.,

$$\pi_k = \pi_0 \prod_{i=1}^{k} \left(\frac{\lambda}{\mu_i} \right), \quad k \in \{1, \ldots, K\}. \tag{3.8}$$

According to the normalization equation,

$$1 = \sum_{i=0}^{K} \pi_i = \left[1 + \sum_{i=1}^{K} \prod_{j=1}^{i} \left(\frac{\lambda}{\mu_j} \right) \right] \pi_0$$

$$= \left[1 + \sum_{i=1}^{k-1} \prod_{j=1}^{i} \left(\frac{\lambda}{\mu_j} \right) + \left(\frac{\lambda}{\mu_k} \right) \sum_{i=k}^{K} \prod_{j=1, j \neq k}^{i} \left(\frac{\lambda}{\mu_j} \right) \right] \pi_0$$

$$= \left(A + \frac{\lambda}{\mu_k} B \right) \pi_0, \tag{3.9}$$

where $A = 1 + \sum_{i=1}^{k-1} \prod_{j=1}^{i} \left(\frac{\lambda}{\mu_j}\right)$ and $B = \sum_{i=k}^{K} \prod_{j=1, j \neq k}^{i} \left(\frac{\lambda}{\mu_j}\right)$. Therefore,

$$\pi_0 = \left(A + \frac{\lambda}{\mu_k} B\right)^{-1}. \tag{3.10}$$

Based on Eqs. (3.8) and (3.10), we have:

$$\pi_K = \left(\prod_{i=1}^{K} \left(\frac{\lambda}{\mu_i}\right)\right) \pi_0 = \left(\frac{\lambda}{\mu_k} \prod_{j=1}^{k-1} \left(\frac{\lambda}{\mu_j}\right) \prod_{j=k+1}^{K} \left(\frac{\lambda}{\mu_j}\right)\right) \pi_0$$

$$= \frac{\lambda}{\mu_k} \prod_{j=1, j \neq k}^{K} \left(\frac{\lambda}{\mu_j}\right) \left(A + \frac{\lambda}{\mu_k} B\right)^{-1}$$

$$= \frac{\lambda \prod_{j=1, j \neq k}^{K} \left(\frac{\lambda}{\mu_j}\right)}{A\mu_k + B\lambda}. \tag{3.11}$$

As λ is a constant, and A and B are independent from μ_k, from Eq. (3.11), we see that when μ_k increases, π_K will decrease monotonically. We thus proved that the capacity increases monotonically with an increasing μ_k. ∎

Proposition 3.1 informs us that with a fixed chain length $K + 1$ and a fixed arrival rate λ, the system capacity ρ_{BDP} is maximized if the service rate for each state is maximized. Applying this proposition to the EFAFS(W, V) model in Fig. 3.2, then with the fixed chain length $\lfloor \frac{M}{W} \rfloor + 1$, ρ_e will increase as V increases, and it will reach its maximum when $V = M$. Note that $V = M$ indicates all channels can be utilized by just one flow.

Proposition 3.2 *A BDP with $K + 1$ states will get its capacity increased if one more state is added to the chain as the last state (i.e., as* State $K + 1$*), and if the service rate of* State $K + 1$ *satisfies $\mu_{K+1} \geq \max\{\mu_k\}, k \in \{1, \ldots, K\}$.* □

Proof We use $\hat{\pi}_k$, $k \in \{0, \ldots, K + 1\}$ to denote the stationary probabilities of the new process after adding *State* $K + 1$. As $\rho_{BDP} = (1 - \pi_K)\lambda$, with λ being a constant, the capacity is inversely proportional to π_K. Therefore, to prove the capacity increases after adding *State* $K + 1$, we are to prove that $\pi_K \geq \hat{\pi}_{K+1}$.

In the old BDP, $\pi_K = \pi_0 \prod_{i=1}^{K} (\lambda/\mu_i)$, and $\pi_0 = \frac{1}{1 + \sum_{i=1}^{K} \prod_{j=1}^{i}(\lambda/\mu_j)}$. When *State* $K + 1$ is appended, the stationary probability of the new state in the new BDP is $\hat{\pi}_{K+1} = \hat{\pi}_0 \prod_{i=1}^{K+1}(\lambda/\mu_i)$, with $\hat{\pi}_0 = \frac{1}{1 + \sum_{i=1}^{K+1} \prod_{j=1}^{i}(\lambda/\mu_j)}$. Therefore,

$$\pi_K \geq \hat{\pi}_{K+1} \iff \pi_0 \prod_{i=1}^{K}(\lambda/\mu_i) \geq \hat{\pi}_0 \prod_{i=1}^{K+1}(\lambda/\mu_i) \tag{3.12}$$

$$\Longleftrightarrow \frac{\prod_{i=1}^{K}(\lambda/\mu_i)}{1+\sum_{i=1}^{K}\prod_{j=1}^{i}(\lambda/\mu_j)} \geq \frac{\prod_{i=1}^{K+1}(\lambda/\mu_i)}{1+\sum_{i=1}^{K+1}\prod_{j=1}^{i}(\lambda/\mu_j)} \tag{3.13}$$

$$\Longleftrightarrow \frac{1}{1+\sum_{i=1}^{K}\prod_{j=1}^{i}(\lambda/\mu_j)} \geq \frac{\frac{\lambda}{\mu_{K+1}}}{1+\sum_{i=1}^{K}\prod_{j=1}^{i}(\lambda/\mu_j)+\prod_{j=1}^{K+1}(\lambda/\mu_j)}. \tag{3.14}$$

To prove $\pi_K \geq \hat{\pi}_{K+1}$ is hence equivalent to proving the inequality of Eq. (3.14).

Firstly, when $\lambda/\mu_{K+1} \leq 1$, Eq. (3.14) holds, as the denominator of the right-hand side is greater than that of the left-hand side.

Secondly, we consider $\lambda/\mu_{K+1} > 1$. Let $P_k = \prod_{i=1}^{k}(\lambda/\mu_i)$, then Eq. (3.14) can be re-written as:

$$\frac{1}{1+\sum_{i=1}^{K}\prod_{j=1}^{i}(\lambda/\mu_j)} \geq \frac{\frac{\lambda}{\mu_{K+1}}}{1+\sum_{i=1}^{K}\prod_{j=1}^{i}(\lambda/\mu_j)+\prod_{j=1}^{K+1}(\lambda/\mu_j)}$$

$$\Longleftrightarrow 1+P_1+\ldots+P_{K+1} \geq \frac{\lambda}{\mu_{K+1}}(1+P_1+\ldots+P_K) \tag{3.15}$$

$$\Longleftrightarrow 1+\left(P_1-\frac{\lambda}{\mu_{K+1}}\right)+\left(P_2-P_1\frac{\lambda}{\mu_{K+1}}\right)+\ldots+\left(P_{K+1}-P_K\frac{\lambda}{\mu_{K+1}}\right) \geq 0. \tag{3.16}$$

Since $\mu_{K+1} \geq \max\{\mu_k\}$, $\forall k \in \{1, \ldots, K\}$, we have $P_1 \geq \lambda/\mu_{K+1}$ and $P_{k+1} \geq P_k(\lambda/\mu_{K+1})$, $\forall k \in \{1, \ldots, K\}$. Hence Eq. (3.16) holds,[6] and Proposition 3.2 is proven. ∎

Proposition 3.2 informs us that a higher system capacity can be achieved by lengthening the CTMC with the service rate of the newly added state being no less than any of the existing service rates in the chain. Indeed, in the strategy EFAFS(W, V), it is true that $\mu_{K+1} \geq \max\{\mu_i\}$, $i \in \{1, \ldots, K\}$. Applying this proposition to the EFAFS(W, V) model described in Fig. 3.2, it indicates that a higher ρ_e can be obtained if the chain is longer, and a longer chain can be achieved by lowering W, i.e., the minimum number of channels for a flow to commence. In other words, in an EFAFS(W, V) system with a fixed maximum number of channels V, the maximum capacity ρ_e is achieved when W is the lowest possible value that satisfies the minimum QoS requirement.

From Proposition 3.2, we can also conclude that blocking new arrivals of flows in order to prioritize ongoing flows does not provide any benefit to the system in terms of maximizing the capacity, because blocking new arrivals reduces the number of

---

[6]Note that $P_{k+1} = P_k\frac{\lambda}{\mu_{K+1}}$, and Eq. (3.16) is in fact equivalent to $1+\left(P_1-\frac{\lambda}{\mu_{K+1}}\right)+\left(P_2-P_1\frac{\lambda}{\mu_{K+1}}\right)+\ldots+\left(P_K-P_{K-1}\frac{\lambda}{\mu_{K+1}}\right) \geq 0$.

ongoing flows in the system, and thus shortens the length of the chain. This explains why Greedy(W, V) can hardly obtain a higher capacity than EFAFS(W, V) under the same system configuration.

Based on Propositions 3.1 and 3.2, we ascertain that the highest system capacity for a BDP is achieved when $V = M$, and the minimum value meeting the requirement of QoS is chosen for W. This is written in formula as:

$$\rho_e = \sum_{k=1}^{\left\lfloor \frac{M}{W} \right\rfloor} \pi_k M \mu = (1 - \pi_{\left\lfloor \frac{M}{W} \right\rfloor})\lambda. \tag{3.17}$$

Though the above maximum capacity is derived for the strategy EFAFS(W, V),[7] in the next section, we will demonstrate that given the same values of W, V, M, λ, and μ, no other strategies can achieve higher capacity than the maximum capacity of EFAFS(W, V).

3.2.4.2 Capacity Upper Bound in General

In an EFAFS(W, V) system, when $V = M$ and W is set to be the minimum possible value, the BDP achieves its maximum capacity as it has the maximum chain length with $\left\lfloor \frac{M}{W} \right\rfloor + 1$ states, and the service rate for every state in this BDP except *State* 0 is at the maximum of $M\mu$. Indeed, the maximum capacity achieved by EFAFS(W, V) is the upper bound for all the strategies that have the same setup for V, M, W, μ, and λ. In order to confirm this statement, in this section, we specifically show that a general strategy has no greater capacity than the capacity upper bound that can be achieved by EFAFS(W, V).

We do this in a two-step manner. In the first step, we transform the states of a general strategy from its own state space, which might be multi-dimensional, to be single dimensional. In other words, we construct an equivalent BDP to the original CTMC, by converting the state space of the original CTMC into a single dimensional one. In the second step, we compare the equivalent BDP with the BDP of EFAFS(W, V). By showing that the chain length of the equivalent BDP cannot be longer than $\left\lfloor \frac{M}{W} \right\rfloor + 1$, and that the corresponding service rates in the equivalent BDP cannot be higher than $M\mu$, we will be able to conclude that the maximum capacity that can be achieved by the EFAFS(W, V) strategy is also the upper bound of capacity for any other strategies.

Firstly, given M and W for a strategy, the number of ongoing flows that can be coexisting in the system is upper-bounded by $\left\lfloor \frac{M}{W} \right\rfloor$. This is equivalent to say that the chain length of the equivalent BDP of the strategy cannot exceed $\left\lfloor \frac{M}{W} \right\rfloor + 1$.

---

[7]The majority of the proof can be found in [6] for CRNs configurations, where only CA is considered.

Recall that the EFAFS(W, V) strategy has its chain length defined as $\lfloor \frac{M}{W} \rfloor + 1$. It is therefore safe to say that given the same system configuration, the chain length of any other strategies will not be longer than that of EFAFS(W, V).

Secondly, we show that the service rates of the equivalent BDP for a general strategy are not greater than $M\mu$.

We denote a state in the general strategy as ϕ, and the stationary state probability of ϕ as $\pi(\phi)$. Note that ϕ can be multi-dimensional, and the number of flows at *State* ϕ is denoted by $|\phi|$. To construct an equivalent BDP with the state space being single dimensional, we re-define the states of the CTMC by an integer pair (r, l), where r is the number of ongoing flows in *State* ϕ, and l is the index of a particular state among all the states that have r flows. *State* (r, l) thus represents the general lth state with r flows. We let $L(r)$ be the number of states that have r flows, and have $l \in \{1, \ldots, L(r)\}$. We also denote by $b(r, l)$ the total number of channels that all ongoing flows utilize at *State* (r, l), and further let $\pi'(r, l)$ be the state probability of (r, l).

Consider that in a general strategy, channel allocation for traffic flows may happen even without flow arrivals or departures. We utilize the parameter $\zeta_r(l, l')$ to represent such transitions. In more details, $\zeta_r(l, l')$ is the transition rate from *State* (r, l) to *State* (r, l'), where $1 \leq l \leq L(r)$, $1 \leq l' \leq L(r)$, and $l \neq l'$.

The balance equation for *State* (r, l) can now be constructed as Eq. (3.18).

$$
\underbrace{(\lambda + b(r,l)\mu)\pi'(r,l)}_{\textcircled{1}} + \underbrace{\pi'(r,l) \sum_{l'=1,\, l'\neq l}^{L(r)} \zeta_r(l, l')}_{\textcircled{2}}
$$

$$
= \underbrace{\sum_{n=1}^{L(r-1)} P_{r-1,n,l}\lambda\pi'(r-1, n)}_{\textcircled{3}} + \underbrace{\sum_{l'=1,\, l'\neq l}^{L(r)} \pi'(r, l')\zeta_r(l', l)}_{\textcircled{4}}
$$

$$
+ \underbrace{\sum_{n=1}^{L(r+1)} P'_{r+1,n,l}b(r+1, n)\mu\pi'(r+1, n)}_{\textcircled{5}}. \tag{3.18}
$$

In Eq. (3.18), $P_{r-1,n,l}$ is the probability of the system moving from *State* $(r-1, n)$ to *State* (r, l) upon a flow arrival, and $P'_{r+1,n,l}$ is the probability that the system moves from *State* $(r+1, n)$ to *State* (r, l) upon a flow departure. $P_{r-1,n,l}$ and $P'_{r+1,n,l}$ represent the different ways of access upon an event in a specific strategy. They satisfy the condition $\sum_{l=1}^{L(r)} P_{r-1,n,l} = 1$ and $\sum_{l=1}^{L(r)} P'_{r+1,n,l} = 1$.

To better interpret Eq. (3.18), we divide it into five parts that are indexed by the circled numbers as shown in this equation. $\textcircled{1}$ and $\textcircled{2}$ represent the transitions out of *State* (r, l), and $\textcircled{3}$-$\textcircled{5}$ describe the transitions into *State* (r, l). When the system reaches the stationary status, ① + ② equals ③ + ④ + ⑤, which is the balance equation for the single *State* (r, l). Specifically speaking, ① involves the transitions out of *State* (r, l) when a flow departure or arrival happens and the number of ongoing flows changes. ③ represents the transitions from other states into *State* (r, l) due to arrival of flows, whereas ⑤ is the transitions into *State* (r, l) because of flows departing. Both transitions in ③ and ⑤ result in a change in the number of ongoing flows. ② and ④ describe the transitions from and into *State* (r, l), respectively, and these transitions do not change the number of ongoing flows in the system.

$$\lambda \sum_{l=1}^{L(r)} \pi'(r, l) + \sum_{l=1}^{L(r)} \pi'(r, l) \sum_{l'=1, \, l' \neq l}^{L(r)} \zeta_r(l, l')$$

$$+ \mu \left(g(r) - \frac{\sum_{l=1}^{L(r)} (g(r) - b(r, l)) \, \pi'(r, l)}{\sum_{l=1}^{L(r)} \pi'(r, l)} \right) \sum_{l=1}^{L(r)} \pi'(r, l)$$

$$= \tag{3.19}$$

$$\lambda \sum_{n=1}^{L(r-1)} \pi'(r - 1, n) + \sum_{l=1}^{L(r)} \sum_{l'=1, \, l' \neq l}^{L(r)} \pi'(r, l') \zeta_r(l', l)$$

$$+ \mu \left(g(r + 1) - \frac{\sum_{l=1}^{L(r+1)} (g(r + 1) - b(r + 1, l)) \, \pi'(r + 1, l)}{\sum_{l=1}^{L(r+1)} \pi'(r + 1, l)} \right) \sum_{l=1}^{L(r+1)} \pi'(r + 1, l).$$

Let $g(r) = \max_l (b(r, l))$, by summing up the balance equations of all states that have r ongoing flows, we have Eq. (3.19).

Equation (3.19) can then be simplified to Eq. (3.21), with $\pi''(r) = \sum_{l=1}^{L(r)} \pi'(r, l)$. Here, we have canceled the equivalent opponents,[8] i.e., $\sum_{l=1}^{L(r)} \pi'(r, l) \sum_{l'=1, \, l' \neq l}^{L(r)} \zeta_r(l, l')$ and $\sum_{l=1}^{L(r)} \sum_{l'=1, \, l' \neq l}^{L(r)} \pi'(r, l') \zeta_r(l', l)$, on both sides of Eq. (3.19).

---

[8]One can see that Eq. (3.20) holds by expanding the summation operation on each side of the equation. The left half of the equation describes the sum for transferring out of all states with r flows. As the flows transferring out of a state will eventually arrive at another state, the left sum equals the sum for going into the destination states with r flows, which is the right half of the equation.

$$\sum_{l=1}^{L(r)} \pi'(r, l) \sum_{l'=1, \, l' \neq l}^{L(r)} \zeta_r(l, l') = \sum_{l=1}^{L(r)} \sum_{l'=1, \, l' \neq l}^{L(r)} \pi'(r, l') \zeta_r(l', l), \tag{3.20}$$

$$\left(\lambda + \mu\left(g(r) - \frac{\sum_{l=1}^{L(r)}(g(r) - b(r,l))\,\pi'(r,l)}{\pi''(r)}\right)\right)\pi''(r)$$

$$=$$ (3.21)

$$\lambda\pi''(r-1) + \mu\left(g(r+1) - \frac{\sum_{l=1}^{L(r+1)}(g(r+1) - b(r+1,l))\,\pi'(r+1,l)}{\pi''(r+1)}\right)\pi''(r+1).$$

Note that Eq. (3.21) has the same format as the balance equation of the rth state in a BDP. The upper half of Eq. (3.21) includes the service rate for states with r services and the lower half contains the service rate for states with the number of services being $r+1$. In the upper half of Eq. (3.21), since $\frac{\sum_{l=1}^{L(r)}(g(r)-b(r,l))\pi'(r,l)}{\pi''(r)} \geq 0$, $\left(g(r) - \frac{\sum_{l=1}^{L(r)}(g(r)-b(r,l))\pi'(r,l)}{\pi''(r)}\right) \leq g(r)$. Similarly, in the lower half of the equation, $\left(g(r+1) - \frac{\sum_{l=1}^{L(r+1)}(g(r+1)-b(r+1,l))\pi'(r+1,l)}{\pi''(r+1)}\right) \leq g(r+1)$. As $g(r) \leq M$ by definition, for any strategies with CA and CF, the corresponding service rates in their equivalent BDPs will be *no greater than* the service rate of $M\mu$, which is the service rate the EFAFS(W,V) strategy can achieve with $V = M$. Recall that the chain length will not be longer than the one in EFAFS(W,V) with W being minimized, according to Propositions 3.1 and 3.2, we ascertain that for any strategy, its capacity will not exceed the capacity defined in Eq. (3.17) for EFAFS(W,V) with $V = M$ and W set to be the smallest possible value determined by the QoS of the flows.

In addition, in Eq. (3.21), if $b(r,l) = g(r)$, $\forall l \in \{1, \ldots, L(r)\}$, we have $\frac{\sum_{l=1}^{L(r)}(g(r)-b(r,l))\pi'(r,l)}{\pi''(r,l)} = 0$, meaning that the maximum service rate, i.e., $M\mu$, is achievable if $b(r,l) = g(r) = M$, $\forall l \in \{1, \ldots, L(r)\}$. This implies that if a strategy meets the following two requirements:

1. the length of the equivalent BDP of the strategy can reach $\lfloor \frac{M}{W} \rfloor + 1$, where W is the minimum value meeting the requirement of QoS,
2. and each of its states with r services is able to utilize M channels, i.e., $b(r,l) = M$, $\forall l \in \{1, \ldots, L(r)\}$ and $\forall r \in \{1, \ldots, \lfloor \frac{M}{W} \rfloor\}$,

then the derived capacity upper bound is *attainable* for the strategy. Dy(0.5, 3) with $M = 3$ is an example of such a strategy, whereas Greedy(0.5, 3) cannot achieve the same upper bound as the chain length is constraint by the strategy.

3.2.5 The Equivalence of EFAFS(W, V) and Dy(W, V)

One may have noticed that the strategies EFAFS(W, V) and Dy(W, V) are very similar to each other in terms of system performance. In fact, Dy(W, V) is equivalent to EFAFS(W, V) for our studied parameters. We now analyze the equivalence by taking a closer look at EFAFS(0.5, 2) and Dy(0.5, 2), with $M = 3$.

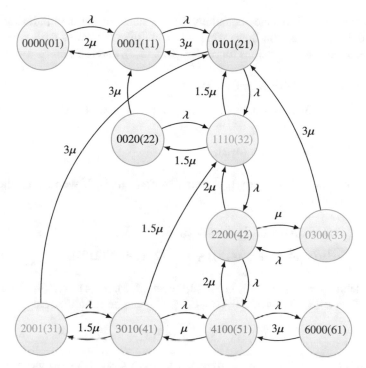

Fig. 3.14 The CTMC of Dy(0.5, 2) when $M = 3$ for elastic flows with index (r, l)

Let us revisit Fig. 3.4 that illustrates the CTMC of Dy(0.5, 2) with $M = 3$ for elastic flows, and use ϕ to represent generally the four-element states in this CTMC. We are to construct a BDP that is equivalent to this CTMC by re-indexing the states according to the number of ongoing flows, r. Then *State* ϕ in the original indexing system will be corresponding to *State* (r, l) (with l being defined the same as the one in Sect. 3.2.4.2) in the new indexing system. Figure 3.14 re-plots the CTMC with states in both indexing systems. The states are in colors with each color representing a group of states that have the same number of flows. For example, *State* (2,1) and *State* (2,2) in the new indexing system correspond to *State* (0,1,0,1) and *State* (0,0,2,0), respectively, in the old indexing system. As both states have $r = 2$, meaning there are two ongoing flows in the system, they are hence represented in the same color of black. Similarly, *States* (3,1), (3,2), and (3,3) represent respectively *States* (2,0,0,1), (1,1,1,0), and (0,3,0,0) for $r = 3$, thus they are all marked in color green. For $r = 4$, the color is brown, and there are *State* (4,1) and *State* (4,2) representing the respective *State* (3,0,1,0) and *State* (2,2,0,0). All the rest states in the old system are also re-indexed in the same way to fit into the new system.

We now establish balance equations for the states in the new indexing system for $Dy(0.5, 2)$ with elastic flows when $M = 3$. Take the states with $r = 3$ as an example, the balance equations for *States* (3,1), (3,2), and (3,3) are:

$$(\lambda + 3\mu)\pi'(3, 1) = 1.5\mu\pi'(4, 1), \tag{3.22}$$

$$(\lambda + 3\mu)\pi'(3, 2) = 1.5\mu\pi'(4, 1) + 2\mu\pi'(4, 2) + \lambda\pi'(2, 1) + \lambda\pi'(2, 2), \tag{3.23}$$

$$(\lambda + 3\mu)\pi'(3, 3) = \mu\pi'(4, 2), \tag{3.24}$$

where $\pi'(r, l)$ is the stationary probability of *State* (r, l). If we sum up the above equations, we have:

$$(\lambda + 3\mu)(\pi'(3, 1) + \pi'(3, 2) + \pi'(3, 3))$$
$$= 3\mu(\pi'(4, 1) + \pi'(4, 2)) + \lambda(\pi'(2, 1) + \pi'(2, 2)). \tag{3.25}$$

Further defining $\pi''(3) = \pi'(3, 1) + \pi'(3, 2) + \pi'(3, 3)$, $\pi''(4) = \pi'(4, 1) + \pi'(4, 2)$, and $\pi''(2) = \pi'(2, 1) + \pi'(2, 2)$, then Eq. (3.25) can be simplified as:

$$(\lambda + 3\mu)\pi''(3) = 3\mu\pi''(4) + \lambda\pi''(2). \tag{3.26}$$

Examining Eq. (3.26) against the BDP of EFAFS(0.5, 2) shown in Fig. 3.2, we can observe that Eq. (3.26) is exactly the balance equation for *State* 3 in EFAFS(0.5, 2). This result applies also to all the other states with the same number of flows. In other words, if we consider the sum of the stationary probabilities of states with the same number of flows r as a single virtual state, the equivalent BDP of the CTMC of $Dy(0.5, 2)$ is in fact the BDP of EFAFS(0.5, 2). As the system parameters we studied in all the previously mentioned strategies are based on the states with the same number of flows, the parameters observed for $Dy(0.5, 2)$ are identical to that of EFAFS(0.5, 2).

To generalize the above example, for a general state (r, l) in the strategy $Dy(W, V)$, its balance equation is:

$$(\lambda + b(r, l)\mu)\pi'(r, l)$$
$$= \sum_{n=1}^{L(r-1)} P_{r-1,n,l}\lambda\pi'(r - 1, n) + \sum_{n=1}^{L(r+1)} P'_{r+1,n,l}b(r + 1, n)\mu\pi'(r + 1, n). \tag{3.27}$$

By summing up all the equations of the states with r flows, we have:

$$\lambda \sum_{l=1}^{L(r)} \pi'(r,l) + \mu g(r) \sum_{l=1}^{L(r)} \pi'(r,l)$$

$$= \lambda \sum_{n=1}^{L(r-1)} \pi'(r-1,n) + \mu g(r+1) \sum_{l=1}^{L(r+1)} \pi'(r+1,l). \tag{3.28}$$

Equation (3.28) holds for $Dy(W, V)$ because $g(r) = b(r, l)$, $\forall l \in \{1, \ldots, L(r)\}$. More concretely, when $r = 1$, $g(r) = 2$, and when $r = 2, \ldots, 6$, $g(r) = 3$ for $Dy(0.5, 2)$ and $M = 3$. We can further define $\pi''(r) = \sum_{l=1}^{L(r)} \pi'(r, l)$, then Eq. (3.28) can be written as:

$$(\lambda + \mu g(r))\pi''(r) \tag{3.29}$$

$$= \lambda \pi''(r-1) + \mu g(r+1)\pi''(r+1).$$

Equation (3.29) is exactly the balance equation of *State r* in the BDP of EFAFS with the same W and V. Therefore, it is not a surprise that strategy $Dy(W, V)$ and strategy $EFAFS(W, V)$ have identical system performance.

3.2.6 System with a Finite Number of Users

In all the previously studied scenarios, we considered systems with an infinite number of users, by assuming λ to be a constant regardless of the number of ongoing flows in the system. In this section we construct CTMC for a scenario where there are a finite number of users in the system.

Consider a system with M channels, and its strategy being $EFAFS(W, V)$ for elastic flows. Suppose there are U users, each of which generates one flow at a time, i.e., one flow per user, as in the traditional telephone system. Each user has an arrival rate λ when it is not being served. In other words, once a user is being served in the system, it will not generate new flows until the service finishes. The service rate for each flow is μ on one channel. To establish a CTMC for the system, let i be the number of users that are being served in the system and thus the state of the CTMC can be represented by i, $i < \min\left(U, \lfloor \frac{M}{W} \rfloor\right)$, where $\lfloor \frac{M}{W} \rfloor$ is the maximum number of ongoing flows in the system. As the system can accommodate at most $\lfloor \frac{M}{W} \rfloor$ flows, and the maximum number of flows that the users can generate is U, the chain length is indeed the minimum of $\lfloor \frac{M}{W} \rfloor$ and U. The transition of the CTMC from *State i* to *State i + 1* is made upon a new flow arriving, and the transition rate is $(U - i)\lambda$. The transition of the CTMC from *State i* to *State i − 1* is made when an ongoing flow finishes and leaves the system. The service rate is $Vi\mu$ when $Vi \leq M$, and $M\mu$ when $Vi > M$.

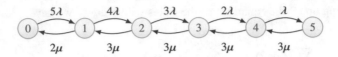

Fig. 3.15 The CTMC of EFAFS(0.5, 2) when $M = 3$ for elastic flows with five users

Figure 3.15 illustrates the CTMC for EFAFS(0.5, 2) with $M = 3$, $U = 5$ for elastic flows. As can be seen from the figure, the only difference between this CTMC and the one for EFAFS(0.5, 2) in a system with an infinite number of users is that the arrival rate of this CTMC decreases when the number of users being served in the system increases.

In the remaining part of the book, we consider only the case where an infinite number of users exist in the system. The main reason is that the one flow per user scenario exists mainly in old fashioned communication systems such as the voice call based system. In modern communication systems, multiple flows can be generated by the same user, resulting in a relatively stable arrival rate of traffic flows at system level, which approximates to an arrival rate that is constant. We thus can use the infinite-user model as a good approximation to most of the communication systems in the scenarios with multiple flows per user.

3.3   Discussions and Summary

In this chapter, we studied CTMC models for different strategies in the single-flow single-user scenario. By examining the system capacity, blocking probability, and the number of channels per flow under different strategies, we revealed the benefit of applying CA and CF for traffic flows. Indeed, CA and CF allow traffic flows to access channels in a more flexible manner, and thus significantly improve system performance in terms of system capacity, blocking probability, and the number of channels per flow.

The scenarios studied in this chapter are designed to be simple. The intention is to let the beginners easily follow the concept and grasp the modeling process of CTMC. In the next chapter, we study a system with multiple types of users, which will be slightly more complicated.

References

1. Jiao L, Li FY, Pla V (2012) Modeling and performance analysis of channel assembling in multichannel cognitive radio networks with spectrum adaptation. IEEE Trans Veh Technol 61(6):2686–2697
2. Jiao L, Balapuwaduge IAM, Li FY, Pla V (2014) On the performance of channel assembling and fragmentation in cognitive radio networks. IEEE Trans Wirel Commun 13(10):5661–5675

3. Jiao L, Li FY (2009) A single radio based channel datarate-aware parallel rendezvous MAC protocol for cognitive radio networks. In: 2009 IEEE 34th conference on local computer networks, pp 392–399
4. Jiao L, Li FY (2009) A dynamic parallel-rendezvous MAC mechanism in multi-rate cognitive radio networks: mechanism design and performance evaluation. J Commun 4(10), 752–765
5. Cordeiro C, Ghosh M (2006) Channel bonding vs. channel aggregation: facts and faction. IEEE 802.22 WG. https://mentor.ieee.org/802.22/dcn/06/22-06-0108-00-0000-bonding-vs-aggregation.ppt
6. Jiao L, Song E, Pla V, Li FY (2013) Capacity upper bound of channel assembling in cognitive radio networks with quasistationary primary user activities. IEEE Trans Veh Technol 62(4):1849–1855

Chapter 4
Markov Chain Analysis of CA and CF with Multiple Types of Users

In the previous chapter, we studied in depth the impact of CA and CF on traffic flows in systems where there exists only one type of users with one type of flows. In this chapter, the influence of CA and CF is studied in a more complicated scenario, i.e., the one in which multiple users exist, and where users have different priorities in using channel resources. CRN is a typical example for such a system, and we study the CRNs where PUs and SUs have one type of flows for each.

In the following sections, we firstly evaluate system performance in the case where the time scales of user flows are similar to each other. We then study the performance of the system in the QSR, which is a special case where the time scale of one type of users greatly exceeds that of the other. Both cases are evaluated by using CTMC analyses. To validate the correctness and the generality of MC analyses, we also employ simulation approaches to examine the performance of the system in general distributions for traffic flows, and this will be studied in the last part of this chapter.

4.1 Analytical Models Based on CTMCs for CRNs

4.1.1 System Configurations and Access Strategies

We model access strategies for elastic flows and real-time flows in the single-flow multi-user case in CRNs. In a CRN, spectrum is allocated to PUs, and SUs are allowed to access the spectrum when PUs are inactive. PUs have higher priority than SUs in terms of spectrum access and can acquire the channels being used by SUs at any time. It is assumed that there is no cooperation between PUs and SUs, thus when a PU flow returns, and when there is no vacant channel for the PU flow to use, one or multiple SU flows must give way to this PU flow.

© The Author(s), under exclusive license to Springer Nature Switzerland AG 2020
L. Jiao, *Channel Aggregation and Fragmentation for Traffic Flows*,
SpringerBriefs in Electrical and Computer Engineering,
https://doi.org/10.1007/978-3-030-33080-4_4

We assume that spectrum sensing[1] carried out by SUs has sufficient accuracy and sensing failures do not influence the statistics of the traffic flows. We also assume that with the help of advanced physical and MAC layer techniques,[2] the statistics of traffic flows will not be influenced by eventual packet drops that are due to, e.g., channel variations. We assume as well that there is a protocol running among the SUs to coordinate CA, CF, and spectrum adaptation, and that the SU flows are independent from each other.

We define the spectrum requirement for a single PU flow to transmit as one channel, and assume the spectrum band in the CRN consists of M channels for PUs. $M \in \mathbb{Z}^+$, and $\mathbb{Z}^+$ denotes the set of positive integer numbers. For the sake of simplicity, we suppose that the system supports CA and CF for SUs, but does not support CA or CF for PUs. Therefore, each PU flow can use only one channel at a time, whereas SUs have an option to transmit a flow using a portion of a channel by CF, or in multiple channels by CA. The multiple channels utilized by an SU flow can be either neighboring to or separated from each other in the spectrum domain. We further define W and V as the minimum and the maximum numbers of channels that a single SU flow can utilize respectively, and let N be the number of channels that an SU flow utilizes, then W, V, $N \in \mathbb{R}^+$, and $0 < W \leq N \leq V \leq M$. $\mathbb{R}^+$ denotes the set of positive real numbers.

As PUs are assumed to occupy exactly one channel for each flow, its channel access strategy is simple. Upon a PU flow arrival, if there is a channel that is not occupied by ongoing PU flows, the new PU flow can be accommodated. Upon a PU flow departure, other ongoing PU flows do not take any actions. For SUs, the case is different. As SUs are enabled with CA and CF, they can adaptively adjust the number of channels that an SU flow utilizes according to channel availability. We assume that SUs use the EFAFS(W, V) strategy that has been introduced in Chap. 3. This means that the ongoing SU flows:

- always utilize as many channels as they are permitted to,
- always *equally* utilize the available channels, and
- always share their occupied channels with a new SU arrival if each of the ongoing SU flows and the new SU flow are able to use at least W channels after channel sharing.

Under the EFAFS(W, V) strategy:

- Whenever there are channel resources being released due to a PU flow or an SU flow completing transmission, or because of an SU flow being forcedly terminated, the released channel resources will be equally shared by ongoing SU flows under the condition that each SU flow is allocated no more than V channels.

---

[1] Spectrum sensing is a technique of CRNs that is designed to prevent interference with PUs and to identify the available spectrum. One can refer to [1] for more details.

[2] A real-time voice conversation may tolerate up to 3% of packet loss [2]. The re-transmission scheme at MAC layer can also reduce the impact of packet loss for traffic flows.

- Upon an SU flow arrival, the new SU flow will commence if the number of channels it can obtain is not lower than W.
- Upon a PU flow arrival, the ongoing SU flows will reduce the number of their occupied channels and possibly be able to continue if at least W channels can be maintained for each SU flow. If the average number of channels is lower than W for the ongoing SU flows after a PU arrives, one or multiple ongoing SU flows will be forced to terminate.

4.1.2 The Precise Model for EFAFS with Elastic Flows

With M, W, and V being defined as in Sect. 4.1.1, we use CTMC to model the EFAFS(W, V) strategy for elastic flows in the CRN that support CA and CF for SUs.

Following the common practice in CTMC modeling, we assume that the arrival of both SU and PU flows follows Poisson process, with an arrival rate λ_S for SU flows and an arrival rate λ_P for PU flows. The service time is assumed to be exponentially distributed with the service rate *per channel* for SU flows being μ_S, and the service rate *per channel* for PU flows being μ_P. Suppose channels are homogeneous, the service rate of an SU flow with N channels is then $N\mu_S$ for elastic flows.

Let i be the number of ongoing PU flows and j be the number of ongoing SU flows in the CRN. The state in the CTMC is then represented by $x = (i, j)$, and the transitions of the chain from a general state (i, j) to any other possible state are shown in Table 4.1. Note that in Table 4.1, we have defined the total number of occupied channels at *State x* as $b(x) = i + \min(M - i, jV)$, and the feasible states of the CTMC as $\mathcal{S} = \{(i, j) | i + Vj < M\} \cup \{x | b(x) = M\}$.

Figure 4.1 shows an example of such a CTMC with $M = 2$, $W = 0.5$, and $V = 1$. In this example, when there is no PU flows existing in the system, the two channels can both be utilized by SUs, and the transitions are depicted by the top branch of the CTMC in this figure. In the top branch, the service rate of all ongoing SU flows is at most $2\mu_S$, as there are in total two channels in this system. The service rate of *State* $(0,1)$ is μ_S because when there is only one SU flow in the system, the SU flow is able to use one channel for transmission, which is the maximum number of channels that is allowed to an SU flow. As ongoing SU flows will share their channels with new arrivals, and the minimum number of channels that is required by an SU flow to commence is 0.5, the system can have at most four ongoing SU flows at the same time, with each of the SU flows utilizing a half of a channel. When there are four SU flows in the system, the CTMC is in *State* $(0,4)$.

Suppose the system is currently in one of the states belonging to the top branch, then upon a PU flow arrival, it will move to a state in the middle branch of the CTMC, where each state has the number of PU flows being 1. For example, if

Table 4.1 Transitions from a generic state $x = (i, j)$ for EFAFS(W, V)

Activity	Destination state	Transition rate	Conditions
PU arrives when no SU exists	$(i + 1, 0)$	λ_P	$i < M$ and $j = 0$
PU arrives when at least one SU exists. No SU forced termination happens	$(i + 1, j)$	λ_P	$M - (i + 1) \geq jW$, $j > 0$, and $i < M$
PU arrives when at least one SU exists. SU forced termination happens	$\left(i + 1, \lfloor \frac{\max(0, M-(i+1))}{W} \rfloor\right)$	λ_P	$M - (i + 1) < jW$, $i < M$, and $j > 0$
PU departure	$(i - 1, j)$	$i\mu_P$	$i > 0$
SU arrival	$(i, j + 1)$	λ_S	$M - i \geq (j + 1)W$
SU departure	$(i, j - 1)$	$\min(M - i, jV)\mu_S$	$j > 0$

In this table, PU and SU represent a PU flow and an SU flow, respectively

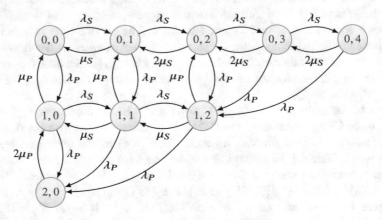

Fig. 4.1 The CTMC of EFAFS(0.5, 1) with $M = 2$ for elastic flows in CRNs

the current state of the system is *State* (0,3) or *State* (0,4), the newly arrived PU flow will terminate one (if it is *State* (0,3)) or two (if it is *State* (0,4)) ongoing SU flows, resulting in the system moving to the destination state, i.e., *State* (1,2). Correspondingly, if the only one PU flow existing in the system leaves, the system will move from the middle branch to a certain state residing in the top branch. Similar analysis can be carried out for the transitions between middle branch and the bottom one.

Table 4.1 summarizes the transitions of EFAFS(W, V) in general. Based on the state transitions presented in Table 4.1, we can construct the transition rate matrix Q, and then calculate the stationary state probability, $\pi(x)$, by solving the equation group $\pi Q = 0$ and $\sum_x \pi(x) = 1$. Note that π is the stationary probability vector of the CTMC.

Our goal is to calculate the following system parameters:

- The capacity of the secondary network, ρ_e. It is calculated by summing up the products of state probabilities and their corresponding SU service rates, as shown in Eq. (4.1):

$$\rho_e = \sum_i \sum_{j=0}^{\left\lfloor \frac{M-i}{W} \right\rfloor} \min(M - i, Vj)\mu_S\pi(i, j).\tag{4.1}$$

In the example illustrated in Fig. 4.1, the capacity is:

$$\rho_e = \mu_S\pi(0, 1) + 2\mu_S\pi(0, 2) + 2\mu_S\pi(0, 3) +$$
$$2\mu_S\pi(0, 4) + \mu_S\pi(1, 1) + \mu_S\pi(1, 2).$$

- The blocking probability, P_{be}, refers to the probability that an SU flow arrival is blocked and lost. It is defined as:

$$P_{be} = \sum_{\substack{x \in S, \\ M-i<(j+1)W}} \pi(x).\tag{4.2}$$

In the example illustrated in Fig. 4.1, the blocking probability of SU flows is the sum of the stationary probability of State (0,4), State (1,2), and State (2,0), i.e.,

$$P_{be} = \pi(0, 4) + \pi(1, 2) + \pi(2, 0).$$

- The forced termination probability, P_{fe}, which is defined as the ratio of the SU flows being terminated over the admitted SU flows, is given by:

$$P_{fe} = \frac{R_{fe}}{\lambda_S^*}$$
$$= \frac{\lambda_P}{\lambda_S^*} \sum_{\substack{x \in S,\ j>0, \\ M-(i+1)<Wj}} \left(j - \left\lfloor \frac{\max(0, M - (i + 1))}{W} \right\rfloor\right)\pi(x),\tag{4.3}$$

where R_{fe} is the forced termination rate, and $\lambda_S^* = (1 - P_{be})\lambda_S$ is the admitted rate that has the blocked SU flows deducted from the total rate of arrival. Indeed, the blocked flows are not included in calculating P_{fe}.

The forced termination probability in the example presented in Fig. 4.1 is:

$$P_{fe} = \frac{\lambda_P(\pi(0, 3) + 2\pi(0, 4) + \pi(1, 1) + 2\pi(1, 2))}{\lambda_S(1 - (\pi(0, 4) + \pi(1, 2) + \pi(2, 0)))}.$$

- The indicator, $\bar{N}_e$, of the number of channels that each SU flow utilizes after CF and CA, is defined as:

$$\bar{N}_e = \frac{E\,(No.\ of\ channels\ utilized)}{E\,(No.\ of\ flows)} = \frac{\sum_{x \in S} \pi(x)(b(x) - i)}{\sum_{x \in S} \pi(x)j}. \qquad (4.4)$$

In the example illustrated in Fig. 4.1, $\bar{N}_e$ can be presented by:

$$\bar{N}_e = \frac{\pi(0,1) + 2\pi(0,2) + 2\pi(0,3) + 2\pi(0,4) + \pi(1,1) + \pi(1,2)}{\pi(0,1) + 2\pi(0,2) + 3\pi(0,3) + 4\pi(0,4) + \pi(1,1) + 2\pi(1,2)}.$$

4.1.3   The Model of EFAFS in the QSR for Elastic Flows

In a CRN, if the time scale between state transitions per PU activity is significantly larger than the time scale between state transitions that are due to SU activities, the CRN is in the QSR. In this subsection, we establish the CTMC model for the EFAFS(W, V) strategy in the QSR and derive the capacity upper bound in the QSR when CA and CF are applied to SUs with elastic flows. In the QSR, PUs appear to be static in comparison with SU events, and the transitions of SU flows reach equilibrium between two consecutive PU events. In other words, if there are i PU flows in the QSR, $i < M$, then the rest $M - i$ channels can be considered as *dedicated* to SU flows. The QSR is valid in scenarios such as communications in TV white band where PU's behavior is relatively stable.

To further explain the concept of the QSR, we plot an example of its CTMC with $M = 2$, $W = 0.5$, and $V = 1$ in Fig. 4.2. As can be seen from the figure, the system models for PU flows and SU flows are described by separate CTMCs. In fact, whether the system is in the QSR or not, the CTMC for PU flows remains

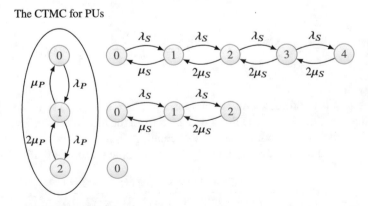

Fig. 4.2 The CTMC of EFAFS(0.5, 1) with $M = 2$ for elastic flows in the QSR in CRNs

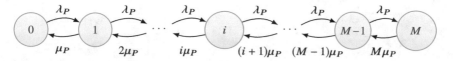

Fig. 4.3 The BDP for PU flows given M channels

the same. This is because PUs are the owners of the spectrum, and the process of PU flows will not be influenced by SU events. A general BDP that describes the transitions per PU flows is shown in Fig. 4.3. Based on the balance equations and the normalization equation of the BDP, the stationary process of the BDP for PU flows can be written as:

$$\pi(0) = \left(\sum_{i=0}^{M} \left(\frac{\lambda_P}{\mu_P} \right)^i \frac{1}{i!} \right)^{-1} , \qquad (4.5)$$

$$\pi(i) = \left(\frac{\lambda_P}{\mu_P} \right)^i \frac{1}{i!} \pi(0), \; \forall i \in \{1, \ldots, M\}.$$

The activities of SUs are subject to the influence of PUs. Therefore, the behavior of SU flows cannot be modeled separately in the same manner without considering PUs' activities. However, in the QSR, by virtue of the time scale of PU flows being significantly greater than that of SU flows, we can construct the model for SU flows by assuming that the available channels are, in a sense, dedicated to SU flows. In the example illustrated in Fig. 4.2, there are three BDPs describing SU flows under different conditions of PU flows. The top branch is when there is no PU flow in the system. In this case, SUs can use up to two channels with maximum four simultaneous ongoing flows, and thus the SU flows can be modeled by the BDP with five states. Similarly, when there is one PU flow in the system, there exists only one available channel for SU flows to use, meaning that the BDP has three states with maximum two simultaneous ongoing SU flows, as shown by the middle branch. The branch at the bottom has only one state, indicating the system does not accommodate any SU flows when there are two PU flows in the system occupying both the channels.

Figure 4.4 generalizes the BDPs for the SU flows in the QSR, with i being the number of ongoing PU flows, and $M - i$ the number of channels being dedicated to SU flows. In Fig. 4.4, we defined $Q = M - i$, $I = \left\lfloor \frac{Q}{W} \right\rfloor$ and $C = \left\lfloor \frac{Q}{V} \right\rfloor$. The service rate of the rth state is $rV\mu_S$ when $0 < r \le C$, and $Q\mu_S$ when $r > C$. This is because when $r > C$, all Q channels are fully utilized by SU flows, and the secondary network reaches its maximum service rate at $Q\mu_S$. However, when $r \le C$, each ongoing SU flow is using V channels, resulting in a service rate of $rV\mu_S$ for the entire secondary system.

The stationary probabilities for the generalized BDP for SU flows can be calculated by solving the balance equations and the normalization equation, and

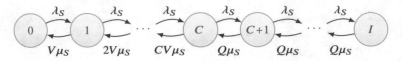

Fig. 4.4 The BDP for elastic SU flows with EFAFS(W, V) given $M - i$ dedicated channels

they can be written as:

$$
\pi(0|i) = \left[\sum_{k=0}^{C} \left(\frac{\lambda_S}{V\mu_S} \right)^k \frac{1}{k!} + \left(\frac{\lambda_S}{V\mu_S} \right)^C \frac{1}{C!} \sum_{k=1}^{I-C} \left(\frac{\lambda_S}{Q\mu_S} \right)^k \right]^{-1},
\tag{4.6}
$$

$$
\pi(j|i) =
\begin{cases}
\left(\frac{\lambda_S}{V\mu_S} \right)^j \frac{1}{j!} \pi(0|i), \ \forall j \in \{1, \ \ldots, \ C\}, \\
\left(\frac{\lambda_S}{Q\mu_S} \right)^{j-C} \frac{1}{C!} \left(\frac{\lambda_C}{V\mu_S} \right)^C \pi(0|i), \ \forall j \in \{C+1, \ \ldots, \ I\},
\end{cases}
$$

where $\pi(j|i)$, $j \in \{0, \ \ldots, \ I\}$ and $i \in \{0, \ \ldots, \ M\}$ denotes the conditional probability of having j SU flows in the system, given that the number of ongoing PU flows is i.

Based on the stationary probabilities calculated in Eqs. (4.5) and (4.6), for the respective primary and secondary networks in the QSR, we can formulate the stationary probability for the entire system as:

$$
\pi(i, j) = \pi(i)\pi(j|i).
\tag{4.7}
$$

Then, the system parameters in the QSR, namely, the capacity of SU flows, the blocking probability of the SU flows, and the number of channels per SU flow can be calculated in the same way as we did in Eqs. (4.1), (4.2), and (4.4). Note that the forced termination probability in QSR is zero as PUs are assumed to be static in this case.

4.1.4   Capacity Upper Bound in the QSR for Elastic Flows

Similar to the single-flow single-user case, for the system of CRN in the QSR, strategies with CA and CF being applied by SU flows under the same system configurations also have an upper bound for the system capacity. The upper bound of capacity is again the one that can be achieved by EFAFS(W, V), with $V = M$ and W set to be the minimum value satisfying the QoS criteria for the elastic SU flows. We derive the upper bound of the capacity in this subsection.

In QSR, PU flows and SU flows can be modeled by separate CTMCs, and the CTMCs for SU flows are composed of SU BDPs under different PU conditions with different numbers of ongoing PU flows. If we can bound the capacity for each

BDP for SU flows, we are able to derive the upper bound of capacity for the overall system by summing up all the individual capacity upper bounds for each of the SU BDPs.

According to Propositions 3.1 and 3.2 in Sect. 3.2.4, in a system with a fixed number of channels, the capacity of the BDP for elastic flows can be improved either by *increasing the service rate of the system* or by *increasing the chain length of the BDP*. In the CRN in QSR with M channels and i ongoing PU flows, the chain length of the SU BDP can be increased with a decreased W, thus a higher capacity can be achieved by setting a smaller value for W. Besides, if V is larger, i.e., each SU flow is allowed to use more channels at the same time, the service rates of the BDP at certain states will be enhanced, resulting in a higher capacity as well. Indeed, the capacity of the BDP is maximized when W is minimized and when $V = M$. If the capacity of each BDP is maximized, the capacity of the entire system is maximized.

As CF can lower W to a degree that is less than one channel, and CA is able to enlarge V by combining fragments of channels, one can observe that in the CRNs in QSR, CA and CF also play an important role in enhancing the capacity of a system.

In Sect. 3.2.4.2, we derived the capacity upper bound for a general strategy with CA and CF in the single-flow single-user case. The method can be also applied to the CRN studied in QSR. The only difference is that in the single-user case, there is only one BDP representing SU activities, whereas in the CRN in QSR, there are more than one SU BDPs, each of which corresponds to a different number of ongoing PU flows in the system, and thus has a different number of available channels being allowed to the SU flows. Fortunately, the difference does not change the nature of the derivation, and we can still use the same method to derive the upper bound of capacity for a general strategy with CA and CF in the CRN studied in the QSR. Indeed, the capacity of any strategies with CA and CF in the CRN in QSR is upperbounded by the maximum capacity that can be achieved by EFAFS(W, V) in QSR with $V = M$ and W being minimized.[3]

Although minimizing the value of W can enhance the capacity of a system, W cannot be infinitely small, as it must satisfy the minimal QoS requirement for an SU flow. Besides, due to the hardware limitation in reality, an infinitely small W is practically unachievable. However, studying the capacity when $W \to 0$ is still of theoretical interests, as when $W \to 0$, the capacity achieved is the firm bound of all the possible capacities that can be achieved in practice.

Proposition 4.1 *With $\pi(i)$ being defined as in Eq. (4.5), then given $W, V \in \mathbb{R}^+$, the theoretical capacity upper bound for elastic SU flows in the CRN in QSR is:*

$$\rho_{qel} = \sum_{i,\ i<M} \pi(i) \min((M-i)\mu_S, \lambda_S). \tag{4.8}$$

---

[3]If we consider queuing as another system resource, the capacity upper bound may be larger as the chain length can be further extended by the queue. Though this is out of scope of this book, one can refer to [3] for a concrete example when queuing is considered.

Proof In a CRN in QSR with $V = M$, when the number of ongoing PU flows is i, $i < M$, the capacity for SUs can be expressed as:

$$\rho' = (1 - \pi(0|i))(M - i)\mu_S. \tag{4.9}$$

To calculate ρ', we only need to calculate $\pi(0|i)$. By solving the balance equations and the normalization equation of the BDP for SU flows with $M - i$ dedicated channels, we have:

$$\pi(0|i) = \begin{cases} \frac{1-r_i}{1-r_i^{I+1}} & \text{if } r_i \neq 1, \\ \frac{1}{I+1} & \text{if } r_i = 1, \end{cases} \tag{4.10}$$

where $r_i = \dfrac{\lambda_S}{(M-i)\mu_S}$. As $I = \left\lfloor \dfrac{M-i}{W} \right\rfloor \to \infty$ when $W \to 0$, we apply limit on Eq. (4.10), and have:

$$\lim_{I \to \infty} \pi(0|i) = \begin{cases} 1 - r_i & \text{if } r_i < 1, \\ 0 & \text{if } r_i \geq 1. \end{cases} \tag{4.11}$$

Substituting Eq. (4.11) into Eq. (4.9) yields:

$$\lim_{I \to \infty} \rho' = \begin{cases} \lambda_S & \text{if } r_i < 1, \\ (M-i)\mu_S & \text{if } r_i \geq 1. \end{cases} \tag{4.12}$$

Equation (4.12) is equivalent to $\rho' = \min(\lambda_S, (M - i)\mu_S)$. As for each value of i, the capacity of SU flows is $\min(\lambda_S, (M - i)\mu_S)$. Considering the probability for each value of i, the capacity of the entire secondary network is $\sum_{i,\, i<M} \pi(i) \min((M - i)\mu_S, \lambda_S)$. Proposition 4.1 is thus proven. □

Proposition 4.1 demonstrates that with $M - i > 0$ dedicated channels, the maximum capacity for the SU BDP equals $\min((M - i)\mu_S, \lambda_S)$. This conclusion agrees with our intuition. When the network is saturated, and if the arrival rate keeps being greater than the total service rate, the number of accomplished flows per time unit is equal to the service rate $(M - i)\mu_S$. Otherwise, the number of flows that can be completed per time unit equals the offered load λ_S. Note that this conclusion holds only for systems in the QSR. The problem of determining the capacity upper bound for SU traffic flows in general cases when CA and CF are enabled is still open.

4.1.5   Models for Real-Time Flows

In this section, we analyze the impact of CA and CF on real-time flows. Since the service time of a real-time flow depends mainly on the user's willingness, and is not affected by the number of adopted channels, the service rate of a real-time flow system is not a function of the number of channels that the SU flows utilize, but a function of the number of ongoing SU flows.

Consider the same EFAFS(W, V) strategy applied to real-time flows. Given the number of ongoing PU flows i, the maximum number of SU flows that the system can accommodate is determined by $\lfloor (M - i)/W \rfloor$. We denote the service rate of the real-time SU flows by μ'_S and the arrival rate of the real-time SU flows by λ'_S. The arrival rate and service rate for PU flows are the same as those defined in the system with elastic flows.

The precise model of a CRN with real-time SU flows has the same transitions as those illustrated in Table 4.1, except that we need to substitute λ_S and $\min(M - i, jV)\mu_S$ with λ'_S and $j\mu'_S$, respectively. As the service rate of a state is a function of the number of ongoing SU flows rather than the number of channels that the flows utilize, the capacity of the system, ρ_r, can be calculated as:

$$\rho_r = \sum_{i, \ i<M} \ \sum_{j=0}^{\lfloor (M-i)/W \rfloor} j\mu'_S \pi(i, j). \tag{4.13}$$

The blocking probability, the forced termination probability, and the number of channels per flow are the same as those in the system with elastic flows, thus can be expressed by Eqs. (4.2), (4.3), and (4.4), respectively.

In the QSR, for any given $M - i > 0$, the behavior of SU flows can be modeled by a BDP with arrival rate λ'_S and the service rate being $m\mu'_S$ for a state with m flows. The chain length of the BDP is $I + 1$, with $I = \lfloor (M - i)/W \rfloor$. In this case, the capacity of the system is:

$$\rho_{qr} = \sum_{i, \ i<M} \pi(i)(1 - \pi(I|i))\lambda'_S$$

$$= \sum_{i, \ i<M} \pi(i) \left[1 - \frac{\lambda'^I_S}{\mu'^I_S I!} \left(\sum_{j=0}^{I} \frac{\lambda'^j_S}{\mu'^j_S j!} \right)^{-1} \right] \lambda'_S. \tag{4.14}$$

In Eq. (4.14), $\pi(I|i)$ is the blocking probability of SU flows given that the number of ongoing PU flows is i. Therefore $1 - \pi(I|i)$ is the probability that a new arrival can be accommodated, and $(1 - \pi(I|i))\lambda'_S$ is the capacity of the ith BDP in the CTMC for SU flows. When we multiply the capacity of the ith BDP with the probability of the corresponding PU event that there are i PU flows existing in the system, the sum of the products, i.e., $\sum_{i, \ i<M} \pi(i)(1 - \pi(I|i))\lambda'_S$, is the capacity of the entire secondary system in the QSR.

4.1.6   Numerical Results

To illustrate the impact of CA and CF on the system performance of the CRN, the numerical results based on our mathematical analyses are presented in this section for both elastic and real-time flows. The system is configured with $\lambda_S = 1$ (λ'_S for real-time flows), $\mu_S = 0.82$ (μ'_S for real-time flows), $\mu_P = 0.5$, and $M = 6$, and we study the system's behavior by observing the system's performance when the arrival rate of PU flows, i.e., λ_P, increases. As λ_P is an indicator of the density of PU flows, this allows us to observe how PU activities influence the performance of SU flows. In terms of strategy configuration, we study EFAFS(1, 1), EFAFS(1, 6), and EFAFS(0.5, 6). One can see that EFAFS(1, 1) is a strategy without CA and CF, EFAFS(0.5, 6) is a strategy where both CA and CF are applied, and EFAFS(1, 6) is considered as a strategy utilizing CA without CF.

We examine the system performance for elastic flows first. Figures 4.5, 4.6, 4.7, and 4.8 illustrate the blocking probability, the capacity, the forced termination probability, and the number of channels per flow, for elastic flows, respectively. In all the three EFAFS strategies, as λ_P grows, PUs become more active, the blocking probabilities and the forced termination probabilities increase, while the capacities decrease. The reason is that when the PUs arrive more frequently, the SU flows have less room for their flows in general. For the same reason, the number of channels per flow for EFAFS(1, 6) and EFAFS(0.5, 6) also decrease as λ_P grows. EFAFS(1, 1) has a constant number of channels per flow, as SU flows in this strategy always occupy one channel for each.

Comparing the different EFAFS strategies with various W and V values, we observe from Figs. 4.5 and 4.6 that the EFAFS(0.5, 6) performs the best with the lowest blocking probability and the highest capacity in general. This manifests the importance of CA and CF in improving the system performance in CRN.

Fig. 4.5 The blocking
probability of elastic flows

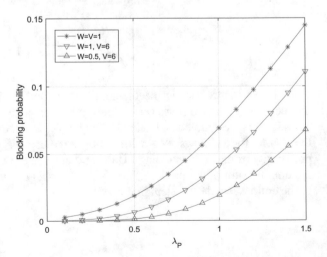

Fig. 4.6 The capacity for elastic flows

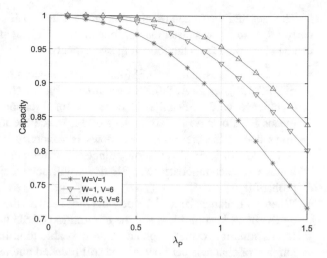

Fig. 4.7 The forced termination probability for elastic flows

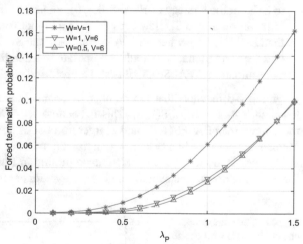

The EFAFS(1, 6) strategy has the second best performance in terms of blocking probability and capacity, indicating that CA without CF also improves the system performance for CRN, though not as much as the strategy with both CA and CF does. The advantage of applying CA and CF becomes more obvious when PUs become more active.

The forced termination probability is the percentage of ongoing SU flows being terminated due to the appearance of PUs. As can be seen from Fig. 4.7, this probability is lower in EFAFS(1, 6) and EFAFS(0.5, 6) than in EFAFS(1, 1). Besides, when λ_P is small, the probability in EFAFS(0.5, 6) is slightly lower than that in EFAFS(1, 6), whereas when λ_P becomes larger, the probability in EFAFS(0.5, 6) becomes closer to EFAFS(1, 6), and surpasses the probability in EFAFS(1, 6) when λ_P increases to a certain value.

The fact that the forced termination probability in EFAFS(0.5, 6) is lower at the early stage, and grows faster than the one in EFAFS(1, 6) after a certain value of λ_P, can be elaborated from the following two aspects:

- When λ_P is small, EFAFS(0.5, 6) can accommodate more flows than EFAFS(1, 6), as EFAFS(0.5, 6) requires only 0.5 channels for an SU flow to commence while EFAFS(1, 6) demands for 1 channel. Besides, a small λ_P means PU flows do not arrive very often, hence the number of terminated SU flows due to the presence of PU flows is generally low for both EFAFS(1, 6) and EFAFS(0.5, 6). Given a low number of terminations in general, the strategy which can accommodate more SU flows will have a slightly lower termination probability.
- When λ_P becomes larger, PUs become more active, the resource that can be used by SU flows becomes less and the termination of SU flows happens more often. In this situation, even though an SU flow is able to utilize multiple channels, the number of channels per flow will be still reduced due to the limited resource left to SUs. In the strategy of EFAFS(0.5, 6), an arrival of one PU flow can result in termination of two SU flows, resulting in a higher forced termination probability than EFAFS(1, 6), where one PU terminates one SU flow. However, considering that a system in practice should not operate in a condition with a high probability of termination, EFAFS(0.5, 6) is still a preferred configuration.

Figure 4.8 displays that the number of channels per flow in EFAFS(0.5, 6) is lower than that in EFAFS(1, 6). This is because EFAFS(0.5, 6) can accommodate more flows than EFAFS(1, 6), and a larger number of flows result in a lower number of channels per flow in a system with a fixed number of channels.

We now study the system performance plotted in Figs. 4.9, 4.10, 4.11, and 4.12 for real-time flows. The overall trend of the system parameters is similar to the case

Fig. 4.8 The number of channels per flow for elastic flows

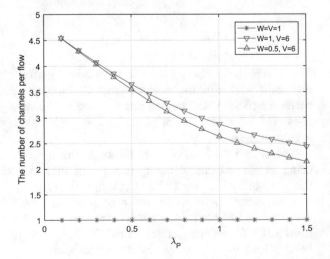

Fig. 4.9 The blocking probability of real-time flows

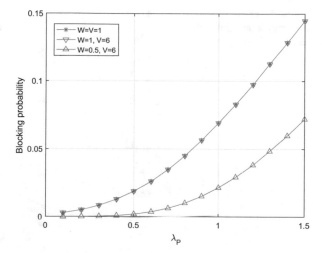

Fig. 4.10 The capacity for real-time flows

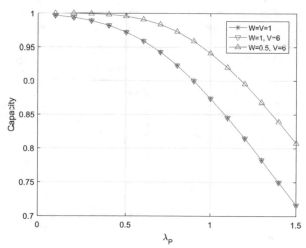

where elastic flows are studied. The most distinct characteristics is that EFAFS(1, 1) and EFAFS(1, 6) perform identically in terms of the capacity, the forced termination probability, and the blocking probability. Indeed, EFAFS(1, 1) and EFAFS(1, 6) in the real-time flow case have identical CTMCs. The reason is twofold. Firstly, both strategies have W set to 1, which means both allow the same maximum number of SU flows in the system, resulting in the chain length being the same. Secondly, although different values are set to V in these two strategies, in the real-time flow case, both strategies have the same service rate, as the service rate is determined by the number of flows in the system, and does not change with V.

Figures 4.9, 4.10, and 4.11 show that EFAFS(0.5, 6) performs better than EFAFS(1, 1) and EFAFS(1, 6) in terms of capacity, forced termination probability, and blocking probability. This is because EFAFS(0.5, 6) can accommodate more SU

Fig. 4.11 The forced termination probability for real-time flows

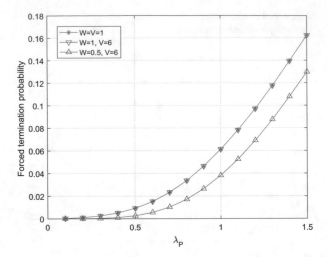

Fig. 4.12 The number of channels per flow for real-time flows

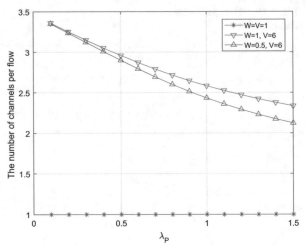

flows in the system with a smaller W, as long as the QoS requirement of the flow is satisfied, and the service rate for a real-time flow is not relevant to V. Therefore, given that the minimum number of channels required for an SU flow to commence meets the lowest QoS requirement, accommodating more flows into a system can indeed improve the system performance statistically for real-time flows.

Figure 4.12 illustrates the indicator for the number of channels per flow. It shows that although CA cannot provide higher service rates for real-time services, it can still increase the number of channels that each ongoing flow utilizes. In this way, the QoS of the real-time flows can be improved by using CA, which is the main benefit that this technique can bring to the real-time traffic systems.

To illustrate the performance of the CRN system in the QSR, we plot the capacity as a function of f in Figs. 4.13 and 4.14. The system is configured with $\lambda_S = 1.5$,

Fig. 4.13 The capacity as a function of f for elastic flows

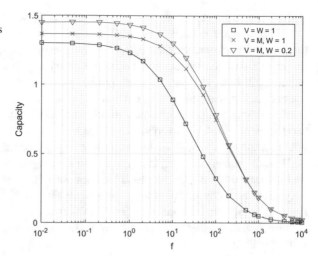

Fig. 4.14 The capacity as a function of f for real-time flows

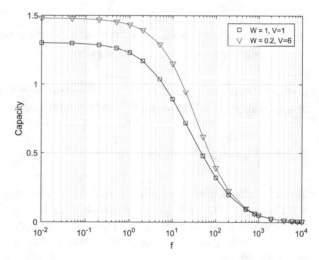

$\mu_S = 0.82$, $\lambda_P = f$, $\mu_P = 0.5f$, and $M = 6$, where f is a scalar describing the dynamics of the primary network. According to Eq. (4.5), $\pi(i)$, $i \in \{0, \ldots, M\}$, is a function of λ_P/μ_P. Given $\lambda_P = f$ and $\mu_P = 0.5f$, $\lambda_P/\mu_P = 2$, implying that the stationary probability of PU flows does not vary with f. Therefore, f can be considered as a parameter that indicates how active PU flows are without changing the stationary distribution of PU flows. A small f means the arrival rate λ_P is low and hence the service rate in the primary network is also low. In this case, PU flows seldom arrive and once they arrive, they stay in the channel for a long time. Correspondingly, if f is large, it means PU flows arrive very often and they stay in the system for only a short period of time. In Figs. 4.13 and 4.14, when f is small, the system operates in a status close to the QSR, whereas when f becomes larger, the behavior of the system is farther away from the QSR.

In the system plotted in Fig. 4.13 for elastic flows, when the system is in the QSR, i.e., when f is small, EFAFS(0.2,6) outperforms EFAFS(1,6), which, in turn, has a better performance than EFAFS(1,1) in terms of capacity. This confirms that the capacity of the system can be improved by applying CA, and can be further enhanced if both CA and CF are applied. Besides, the theoretic upper bound of the capacity in the QSR can be calculated by applying Eq. (4.8). Given the specific configurations of the system, this value is calculated to be 1.4572. Inspecting the capacities in Fig. 4.13, we observe that the EFAFS(0.2,6) strategy in the QSR is able to achieve a capacity that is very close to the theoretical value.

When f becomes larger, the system will no longer stay in the QSR. We observe that the capacity decreases until it reaches zero. This is because PU flows arrive very often and stay in the system for a short time, which, from SUs' perspective, is an interruption of their ongoing flows and results in a high probability of termination. The capacity of the system is degraded because of the high probability of termination.

4.2 Simulations and System Performance with Various Measurement-Based Distributions

So far, we have employed CTMC to model the system behavior mathematically. In this section, we will validate the CTMC models through computer simulations.

4.2.1 Simulation Procedure

The standard method to simulate a CTMC is presented in [4]. The concept is to simulate the evolution of the system by a discrete sequence of events along time, and observe the events statistically. Algorithm 1 describes the simulation procedure. In this method, an instantaneous state at time t, which is usually multi-dimensional, is denoted by $x(t) = (x_1(t), \ldots, x_i(t), \ldots, x_n(t))$, where $x_i(t)$, $i \in \{1, \ldots, n\}$ is a scalar representing the ith element in $x(t)$. It is assumed that there are l possible destination states upon a transition from state $x(t)$, with the transition rate associated with the kth destination state being represented by $\xi_k(x(t))$, $k = 1, \ldots, l$. τ is the duration of time between two consecutive transitions.

To better understand the simulation procedure, we simulate the BDP in Fig. 4.15 as an example. The BDP in Fig. 4.15 is instantiated from the general BDP for PU flows in Fig. 4.3 with $M = 3$. In this instantiated BDP, the state of the system at time t is the number of PU flows at that time, and is represented by a one-digit element in $x(t)$ as $x(t) = (x_1(t))$. The system has four possible states, i.e., $\{x\} = \{(0), (1), (2), (3)\}$.

We simulate the BDP in the following steps:

1. At the beginning of the simulation, without loss of generality, we assume there are two flows in the system. Therefore, the state of the system is (2), i.e., $x(0) = (2)$.

Algorithm 1 Simulation-CTMC

Input The set of possible states $\{x\}$, the set of transition rates for all states $\{\xi_k(x)\}$, and simulation duration t_s.

Output The statistics of interested parameters.

1: Select an initial state of the system, $x(0)$.
2: **while** $0 \le t \le t_s$ **do**
3:    For the current state $x(t)$ of the system, find its transition rates $\xi_k(x(t))$, $k = 1, \ldots, l$.
4:    Calculate the sum of all transition rates for the state, $\xi(x(t)) = \sum_{k=1}^{l} \xi_k(x(t))$.
5:    Generate the duration of time, τ, by sampling from an exponential distribution with mean $1/\xi(x(t))$.
6:    Determine the destination state $x(t + \tau)$. The kth destination state is selected with probability $\xi_k(x(t))/\xi(x(t))$ by sampling from the discrete distribution $\xi_k(x(t))/\xi(x(t))$, $\forall k = 1, \ldots, l$.
7:    Update the new time $t = t + \tau$, the new system state $x(t + \tau)$, and the parameters of interest.
8: **end while**
9: Calculate the statistics of the parameters of interest.

Fig. 4.15 The BDP for PU flows given $M = 3$ channels

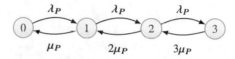

2. From *State* (2), the system can either move to *State* (3) with rate λ_P, or to *State* (1) with rate $2\mu_P$. Therefore, $\xi(x(0)) = \sum_k \xi_k(x(t)) = \lambda_P + 2\mu_P$.
3. We determine the transition time τ by sampling from an exponential distribution with mean $1/(\lambda_P + 2\mu_P)$.
4. We decide the destination state $x(0 + \tau)$ by referring to a random number drawn from the uniform distribution in [0,1]. If the random number is between 0 and $\lambda_P/(\lambda_P + 2\mu_P)$, the destination state is *State* (3). Otherwise, the system goes to *State* (1).
5. We increase the index of the system time by τ, update the system state to the new destination state, and register the events occurred during the period of time τ.

This process is repeated until the end of the given duration of time, t_s.

If we want to simulate a system parameter, e.g., the capacity, we can count the number of completed PU flows in the process of simulation, and divide the number by the period of time at the end of the simulation. The total number of completed PU flows in the simulation can be simply obtained by registering each PU flow's completion, i.e., whenever the system transfers to a destination state with one less ongoing PU flows.

The Matlab code for the above simulation procedure is presented in Listing 4.1, where the capacity is simulated as an example of simulating the system parameters of interest.

To illustrate the results of simulations, we plot in Fig. 4.16 the analytical and simulated capacities of the EFAFS strategy as a function of λ_P for elastic flows. The system is configured with $\lambda_S = 1.5$, $\mu_S = 0.82$, $\mu_P = 0.5$, and $M = 6$.

In this figure, the curved lines represent the analytical results, and the triangle and square marks are results obtained from simulations. From the figure, we observe that the simulated capacities coincide with the analytical ones, implying the simulation is precise, and is a good approximation to the analytical results. Indeed, if the simulation is carried out for a longer period of time, even more precise results can be expected from the simulations.

```matlab
1   clc ; clear all ; close all ;
2   Lambda_P=0.4; %Flow arrival rate
3   Mu_P=0.5; %Flow departure rate
4   M=3; %Number of available channels
5   T_s=10000000; %The time of the simulation
6   X_0=0;%Intitial state of the simulation
7   Index=1;%The index of transitions
8   X(Index)=X_0;%Register the number of flows (State of the CTMC)
9   T(Index)=0;%Regester the accumulated durations
10  Completed=0;%Regester the number of completed flows
11  while T(Index)<T_s %When the time is less than the simulation
        time
12      x=X(Index);%The current state
13          if 0<=x && x<M %If the state is not the one with M flows
14              tau=exprnd(1/(Lambda_P+Mu_P*x));%Generate the interval
                    before the next transition
15              T(Index+1)=T(Index)+tau;%Add the generated interval to
                    the accumulated system time
16              u=rand; %Generate a number uniformly in (0,1)
17               if u<((x*Mu_P)/(Lambda_P+x*Mu_P))%Determine the
                    destination state
18                   X(Index+1)=x-1; %One flow departs , and update
                        system state
19                   Completed=Completed+1;%Update the total number of
                        completed flows
20              else
21                   X(Index+1)=x+1;%One flow arrives
22              end
23          elseif x==M %If the state is the one with M flows
24              tau=exprnd(1/(x*Mu_P));%Generate the interval before
                    the next transition
25              T(Index+1)=T(Index)+tau;%Add the generated interval
                    to the system time
26              X(Index+1)=x-1;%One flow departs , and update system
                    state
27              Completed=Completed+1;%Update the total number of
                    completed flows
28          else
29              break;
30          end
31              Index=Index+1; %Update the index
32  end
33  Capacity=Completed/t_s; %Capacity is the completed flows over the
        total simulated durations
```

Listing 4.1 Matlab code for simulation: the BDP for PU flows with $M = 3$

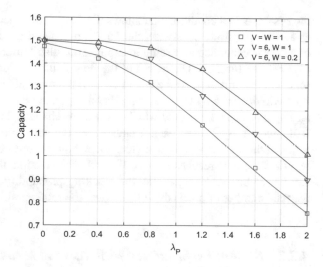

Fig. 4.16 The capacity of EFAFS for elastic flows with simulation results

Another way of doing simulation for a CTMC is to track the system along time by dividing the time into sufficiently small intervals and check the events of the process, namely, the arrival and departure of a flow, within each interval. Specifically speaking, the simulation is conducted over a period of time with N time units, with each unit being further divided into n sub-intervals. Denoting the time unit as T, and the sub-interval as h, then h is the minimum time granularity in this simulation, and $T = nh$. Note that h is to be configured sufficiently small so that the probability of simultaneous events in one sub-interval is ignorable. When the simulation is carried out, within each sub-interval h, the arrival or departure event of the PU or SU flows are checked, and processed according to the strategy that the system adopts. For the Poisson arrival process, arrivals of a certain flow within each sub-interval follow the binomial distribution[4] with parameters $\frac{\lambda_i T}{n}$, where λ_i is the arrival rate of the traffic flows for the ith type of users. The lengths of flows are generated according to exponential distributions with parameters corresponding to the types of users. At the end of each h, the length of the ongoing flow decreases by a certain number, depending on how much of the flow is completed. The flow is considered as completed when the length of the flow reaches zero. At the end of simulation, the statistics of the system can be calculated.

We can explain the above simulation procedure by an example where the strategy EFAFS(W, V) is adopted. The arrivals of PU and SU flows in each sub-interval h are generated according to the binomial distributions with parameters $\lambda_P T/n$ and $\lambda_S T/n$, respectively. For example, when the binomial distribution of SU indicates an SU arrival, the simulator will check the possibility to accommodate the flow. If

---

[4]Note that the binomial distribution converges to a Poisson distribution when $n \to \infty$, i.e., $\lim_{n \to \infty} \frac{n!}{k!(n-k)!} (\frac{\iota T}{n})^k (1 - \frac{\iota T}{n})^{(n-k)} = e^{-\iota T} \frac{(\iota T)^k}{k!}$. In our simulations, as T is the time unit and n is sufficiently large, the probability $\frac{\iota T}{n}$ is sufficiently small.

the system is in a condition that allows the flow to commence, the flow is registered as an ongoing flow. Otherwise, the flow is registered as being blocked. The length of a flow is generated according to exponential distributions, with parameter μ_P for PUs and μ_S for SUs. The length of the flow is then divided into segments that are scaled to the unit of h. At the end of each h, the remaining length of an ongoing PU flow will decrease by one, meaning that a segment of this flow has completed. For an ongoing elastic SU flow which utilizes N channels simultaneously, the length will decrease by N after each h ends, since the service rate for this SU flow is N times of the service rate of one channel. A flow is considered as finished when the length of the flow reaches zero. At the end of simulation, the statistics of the system can be calculated. For example, the blocking probability is obtained by averaging the total number of blocked SU flows over the total number of arrived SU flows.

4.2.2   System Performance with Various Measurement-Based Distributions

When we use CTMC to model a system, and when we employ the simulation methods introduced in Sect. 4.2.1 to simulate a system, we utilize the distribution with the memoryless feature. However, in reality, the distributions of traffic flows may not have the memoryless property, and must be simulated according to distributions based on real-life measurements. Briefly speaking, instead of assuming that the arrival events follow Poisson distributions, and that the flow lengths are exponentially distributed, in the simulation for a non-memoryless system, both the duration of time between two arrival events for a certain type of flows and the length of the flow are generated by certain random variable generators. Those random variable generators are normally constructed according to measurement results of real communication systems, and the resulting distributions are called measurement-based. The statistics of the system obtained by applying the measurement-based arrivals and flow lengths can help us understand to what extent the results based on CTMC models reflect the system's performance in real-life.

Figure 4.17 illustrates the blocking probability of the EFAFS(0.2, 6) strategy applied to elastic traffic with $M = 6$. The analytical results from the CTMC model are depicted and labeled as *Original* as a reference, while the other curves are all based on simulations. We consider three simulation cases with different traffic patterns [5, 6]. In the case *Randomwalk*, the behavior of PUs is generated by a random walk model[5] [5], the arrival of SU flows follows a Poisson process, and a log-normal distribution is used to simulate flow lengths. In the cases of *Lognormal* and *Lognormal**, Poisson processes are utilized to model the arrivals of PU and SU flows, and the flow lengths are generated according to log-normal distributions.

---

[5]Note that the random walk is adopted to describe the presence of PU flows, rather than the PUs' movement behavior.

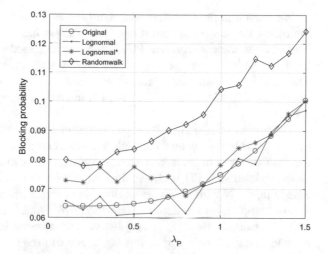

Fig. 4.17 The simulation results of blocking probability of EFAFS for elastic flows with various measurement-based distributions

However, in Case *Lognormal*, the means and the variances of the log-normal distributions are equal to those of the corresponding exponential distributions in the CTMC, while in Case *Lognormal**, the means are the same, but the variances are greater than those of the corresponding exponential distributions. The squared coefficient of variation (SCV) for Case *Lognormal** is configured as SCV=4.618 [6] (SCV= variance/mean^2).

The results in Fig. 4.17 indicate that when the simulation of the system does not maintain the memoryless property for some of the distributions, the results will differ away from the one that is obtained by the CTMC analytical model. However, the difference is not significant. One can observe from the figure that the results of the two log-normal cases are in fact fairly close to the analytical one, with the difference being within 1.5% for most of the values of λ_P. This confirms that the results based on CTMC can represent the system's performance as long as flow arrivals are Poisson processes. The results obtained in the random walk case is about 3% larger than the analytical results, which exhibit, though not precisely, a rough estimate of the system performance. All in all, the CTMC model of a system can be considered as a reference in analyzing the system performance for traffic flows with CA and CF.

4.3 Discussions and Summary

In this chapter, we studied the CTMC models for CRNs with CA and CF applied to SUs, and introduced approaches to simulating the CRNs in order to validate the CTMC models. The analytical results show that the system performance can be significantly improved by applying CA and CF to allow traffic flows to access the channels in a more flexible manner, and CTMC models are in general good approximations to the systems of CRNs.

We claim that the CTMC models built for the CRNs can also be utilized to the multi-flow single-user system if one type of flows has firm priority over the other. Then the parameters represent PUs in the previous model can be used to describe the flows with the firm priority in the new system. Indeed, CTMC, as a general analytical tool for stochastic process, can be utilized to describe distinct systems, as long as the CTMC can reflect the nature of interactions among the flows or the users precisely.

The scenario studied in this chapter is a simple example for the single-flow multi-user system. One can extend the study to a more complicated case, where the system has multiple types of users, and each type of users have multiple types of flows with various priorities. In [7], such a multi-flow multi-user case is studied by configuring SUs in the CRN with two types of flows in different priorities. Regarding EFAFS, a more complicated scenario is examined in [8], where a PU flow can also utilize multiple channels. Both [7] and [8] analyzed the systems in the QSR also. Although queuing is out of the scope of this book, one can refer to [3] to see that the same modeling and simulation procedure can be followed by modeling with queuing, and the performance of a system with queuing can be analyzed in the same way as presented in this chapter. It is also worth mentioning that to support CA and CF, a system needs to have more complicated signaling process. The signaling overhead is studied analytically in [9]. The readers who have reached so far are expected to be able to understand thoroughly these research articles.

References

1. Yucek T, Arslan H (2009) A survey of spectrum sensing algorithms for cognitive radio applications. IEEE Commun Surv Tutorials 11(1):116–130
2. Recommendation, G (2001) 1010 End-user multimedia QoS categories. ITU-T, November
3. Balapuwaduge IAM, Jiao L, Pla V, Li FY (2014) Channel assembling with priority-based queues in cognitive radio networks: strategies and performance evaluation. IEEE Trans Wirel Commun 13(2):630–645
4. Gillespie DT (1976) A general method for numerically simulating the stochastic time evolution of coupled chemical reactions. J Comput Phys 22(4):403–434
5. Willkomm D, Machiraju S, Bolot J, Wolisz A (2009) Primary user behavior in cellular networks and implications for dynamic spectrum access. IEEE Commun Mag 47(3):88–95
6. Barford P, Crovella M (1998) Generating representative web workloads for network and server performance evaluation. In: Proceedings of the 1998 ACM SIGMETRICS joint international conference on measurement and modeling of computer systems, SIGMETRICS '98/PERFOR-MANCE '98. ACM, New York, pp 151–160
7. Jiao L, Li FY, Pla V (2012) Modeling and performance analysis of channel assembling in multichannel cognitive radio networks with spectrum adaptation. IEEE Trans Veh Technol 61(6):2686–2697
8. Jiao L, Balapuwaduge IAM, Li FY, Pla V (2014) On the performance of channel assembling and fragmentation in cognitive radio networks. IEEE Trans Wirel Commun 13(10):5661–5675
9. Balapuwaduge IAM, Jiao L, Li FY (2012) Complexity analysis of spectrum access strategies with channel aggregation in CR networks. In: 2012 IEEE global communications conference (GLOBECOM), pp 1295–1301

Chapter 5
Test-Bed Evaluation of CA and CF via a Software Defined Radio

In Chaps. 3 and 4, we presented the analytical models and simulation approaches to study the impact of CA and CF on traffic flows in the single-flow single-user and the single-flow multi-user systems. In this chapter, we investigate the impact of CA and CF in a test-bed. We employ a software defined radio (SDR) from National Instruments (NI) to evaluate the performance of a CR system with user datagram protocol (UDP) flows. The adopted SDR is based on LTE protocol stack with additional functionalities that can support CA and CF, and accommodating PUs. By conducting measurements based on the test-bed system, we will be able to confirm that performance improvement can indeed be obtained by applying CA and CF in a real system.

In the following sections, we firstly introduce briefly the basic configurations of the system and the strategies that are to be evaluated. We then explain in detail the hardware setup and the workflow of the system. The numerical results based on the measurements are illustrated at the end of this chapter. An abridged version of this chapter is published in [1].

5.1 The Communication Scenario and the Access Strategies

In the test-bed CRN, two CRs compose one SU communication pair, with one of them being the transmitter and the other being the receiver. We consider two system configurations: one is with CA and CF, and the other is without. When CA and CF is configured, the channel accessing strategy adopted in the CRN is EFAFS(W, V), with W being the minimum number of channels required for a flow to start transmission, and V being the maximum number of channels that a flow can occupy simultaneously. In the EFAFS(W, V) system, the ongoing SU flows share equally the available spectrum with each other. Newly arrived SU flows are allowed to start transmitting by sharing the channels with ongoing SU flows, as long as the

L. Jiao, *Channel Aggregation and Fragmentation for Traffic Flows*,
SpringerBriefs in Electrical and Computer Engineering,
https://doi.org/10.1007/978-3-030-33080-4_5

number of channels that each SU flow occupies after channel sharing is greater than or equal to W. If a PU flow arrives, and if there is no vacant channel for the PU flow, the SU reduces the amount of channels shared by each flow or even terminate one or more ongoing SU flows to gather the channel resources that the PU flow requires. If a PU or an SU flow departs, the released channel resource can be shared by ongoing SU flows.

To set a benchmark for the test-bed measurement, we also do a series of measurement in the test-bed CRN without CA and CF. In the configuration without CA and CF, each SU flow occupies only one channel for communication. If PU comes to a channel that is in-use by an SU flow, the SU flow will move to another channel to continue with communication if there is at least one free channel. If all the channels are being occupied, the SU flow will be terminated upon the PU's arrival.

We specifically select UDP flows as the type of traffic flows being used by SUs in the test-bed CRN. This is because the UDP protocol is connectionless, hence allows the measurement results to neutrally reflect the impact of the CA and CF. If we utilize connection oriented protocols, for example, the transmission control protocol (TCP), then the system performance will be influenced by the congestion control functionality of the TCP, making the observation results for CA and CF biased. Besides, under the TCP, a flow may be re-transmitted for several times when it is terminated, and this will result in a more complicated statistics of the SU flows. In addition, the acknowledgment (ACK) message of TCP requiring a feedback communication link brings in another uncertainty in scenarios where the ACK message can be eventually interrupted by PUs. Therefore, considering both the simplicity and the neutrality of observation results, UDP flow is the best option that we can use in the test-bed CRN.

5.2 System Design

5.2.1 Test-Bed Setup

The test-bed is based on the NI platform, which consists of a software platform named LabView and a hardware platform that can be formed from different types of specialized hardware. In this study, we choose the NI-USRP2953R to form our radio platform. NI-USRP2953R is an SDR and belongs to the category called Universal Software Radio Peripheral (USRP). There are indeed many research activities that have been carried out on LabView and USRP, and one can refer to [2] and [3] for two of them.

Our test-bed is specifically made up from a USRP, the LabView Communications Design Suit 2.0, and the LTE Application Framework developed by NI. Figure 5.1 shows the setup. The USRP is utilized as the transmitter (Tx) and the receiver (Rx), and the transmissions between Tx and Rx are via an over-the-air interface. In

Fig. 5.1 A photo of the test-bed

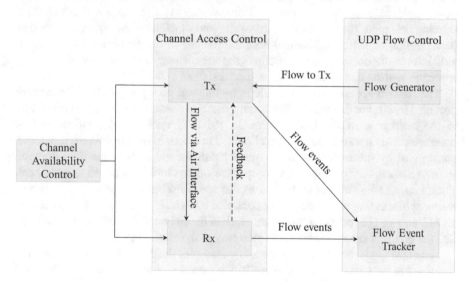

Fig. 5.2 The block diagram of the CRN test-bed

other words, flows are sent by the Tx through the transmission antenna, and the Rx receives flows through the receiving antenna. The functionality of the USRP is controlled by the controller in collaboration with the LabView software. The controller is composed of the PXIe-8880 and the PXIe-1082 (or PXIe-1071), which are specialized hardware that supports real-time deterministic control of the SDR.

Figure 5.2 illustrates the main modules of the test-bed, and they are the channel availability control module, the channel access control module in collaboration with the LTE framework, and the UDP flow control module. The three modules

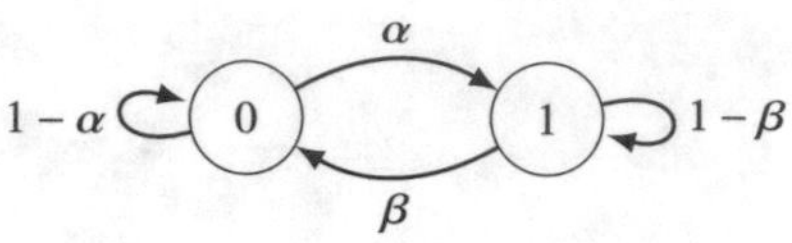

Fig. 5.3 The DTMC of the
PUs

$$1 - \alpha \; \overset{}{\underset{}{\bigcirc}} \; \boxed{0} \; \overset{\alpha}{\underset{\beta}{\rightleftarrows}} \; \boxed{1} \; \overset{}{\underset{}{\bigcirc}} \; 1 - \beta$$

form the logic control of the test-bed, and are made in the software LabView and implemented on the controller in order to control the USRP interface.

The channel availability control module determines whether a PU is active on one channel, and the PU's status, in turn, determines whether this channel is available for an SU flow. For each channel, the module maintains a two-state DTMC to generate the active/inactive sequence that describes the PU's activity. As shown in Fig. 5.3, in the DTMC, *State* "0" means the PU is inactive, implying that the channel is idle and available for SUs to use. *State* "1" means there is an active PU in this channel and SUs must not utilize the channel when the PU is active. Indeed, PUs operate in a time-slotted manner, and the statistics of PU activities on each channel is independent and identical, as each channel is controlled by an independent DTMC. In principle, SUs need to carry out spectrum sensing to determine whether a channel is being occupied. However, to eliminate the impact of imperfect spectrum sensing on UDP flow statistics, in this test-bed, the module of channel availability control will inform the channel access module the state changes of the channel, so that spectrum sensing can be considered as perfect.

The channel access control module is established on top of the LTE framework provided by NI. For this reason, the communication channel is configured according to the realization of the LTE system. In the LTE system, the smallest unit of communication resource that can be allocated to a certain flow is the resource block constructed by subcarriers in the frequency domain and slots in the time domain. In this test-bed, the spectrum of a group of resource blocks is defined as a channel. Therefore, an idle channel is defined as the spectrum band covered by a group of resource blocks that are currently not occupied by PUs. If a channel is being used by an SU, and a PU appears on this channel, then the SU needs to release the channel, i.e., the group of resource blocks, to give way to the eligible PU.

Note that the feedback information, for example, the ARQ messages, which are supposed to be sent from the receiver to the transmitter at MAC layer, is handled internally in the USRP in this test-bed. In other words, we do not establish another air interface between Tx and Rx to transmit the feedback information. Instead, the USRP handles the information internally and ideally so that the transmission of feedback information can be considered perfect. For this reason, the arrow for the feedback information in Fig. 5.2 is presented by a dashed line.

As the EFAFS strategy requires all available channels to be equally shared by ongoing UDP flows, in the test-bed, we apply the weighted fair queuing (WFQ) for UDP flows to achieve uniform access opportunities at packet level. Specifically, depending on the configuration of the EFAFS, the packets that are sent to the USRP via localhost are transmitted over-the-air interface [4] on a certain number of available channels based on WFQ. In the benchmark configuration where CA and CF are disabled, only one idle channel is utilized for transmitting packets from one UDP flow.

The UDP flow control module contains two components. One is the Flow Generator, and the other is the Flow Event Tracker. The Flow Generator is responsible for generating UDP flows. As the test-bed is a time-slotted system, we suppose the arrival of the UDP flows follows a Bernoulli distribution with probability P_a, and the size of the flows, X, follows a geometric distribution with parameter P_s, i.e., $Pr(X = k) = (1 - P_s)^{k-1} P_s$, for $k = 1, 2, \ldots$. Then at the beginning of each time slot, an SU flow arrives with probability P_a. If there is an SU flow arriving, and if there are channels available for the SU flow, the Flow Generator generates a flow with the size that follows the geometric distribution $Pr(X = k)$.

The Flow Event Tracker registers flow events, maintains the flow statistics, and calculates numerical results.

Figure 5.4 illustrates the workflow of the test-bed within one time slot. At the beginning of each time slot, the system starts from S_1, where it checks the status of the PUs and SUs, including the number of ongoing SU flows, the number of channels that the SU flows occupy, the arrivals and departures of PUs, and the number of channels that the PUs occupy. Then based on the channel access rules at S_2, the system checks at S_3 to verify if the number of channels in this system is sufficient to support all ongoing flows or not. If the number is insufficient, the system makes a decision to go to S_4 to terminate one or more SU flows and register the termination events. Otherwise, it goes to S_5 to check if there are new SU flows arriving. If there are no new SU flows arriving, the system goes to S_8. Otherwise, it goes to S_6 to verify if there are sufficient number of channels that can be used to accommodate the new SU flow. If the number of available channels plus the number of channels that can be shared by ongoing SU flows is insufficient to accommodate the new SU flow, the new SU flow will be blocked, and the system goes to S_7 to register a blocking event. If the channel condition meets the requirement for an SU flow to commence, the system will accommodate the new SU flow and goes to S_8. S_8 is a state where all the ongoing flows are being transmitted over the channels. At the end of the time slot, the system moves to S_9 and register the number of the successfully transmitted flows within this slot.

The above process repeats in every time slot during the period of measurement.

5.2.2 Performance Metrics

Similar to the metrics utilized in the previous analytical models and the simulation approaches, we study the system performance of the test-bed by examining the percentage of SU flows that are blocked, terminated, and completed during the measurement, and they are denoted by P_b, P_{fd}, and ρ_{pc}, respectively. A flow is blocked when it arrives at the system but is prevented from entering the system for communication. A terminated flow is an ongoing flow that is terminated due to the appearance of the PUs. A flow is completed when it is successfully transmitted over

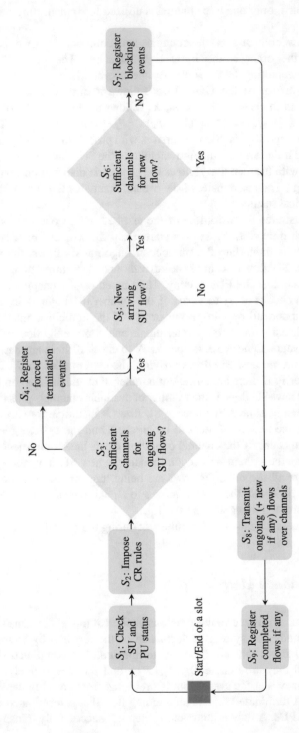

Fig. 5.4 The workflow for each slot of the CRN test-bed

the communication link. These three system parameters correspond respectively to the blocking probability, the forced termination probability, and the capacity studied in the analytical models and the simulation approaches. The difference is that the unit of the capacity applied in the analytical models and simulations is flow per time unit, whereas in the test-bed, we normalize the capacity by the injected density of UDP flows. Therefore, the capacity of the test-bed is described in percentage and we have[1] $(1 - P_b)(1 - P_{fd}) = \rho_{pc}$.

5.3 Measurement Results

5.3.1 Numerical Configurations of the Test-Bed

In this test-bed, the system operates on a spectrum band centered at 2 GHz with a total bandwidth of 20 MHz. The 20 MHz spectrum band is divided into 25 resource block groups, each of which covers 0.8 MHz. Each resource group is considered as an independent channel with a PU activity described by the DTMC in Fig. 5.3. The length of each time slot is 250 ms, and P_a and P_s are configured as 0.4 and 0.6 respectively. The unit of k is 0.1 mega bytes, i.e., the flow length is a multiple of 0.1 mega bytes. The parameter β in the DTMC that describes the transitions from occupied to idle, is configured as 0.75. In the Strategy EFAFS(W, V), we set $W = 0.5$ and $V = 25$, meaning that one flow requires at least half of a channel and can occupy up to 25 channels. The measurements are done for ten times in total in the same indoor and static environment, and are carried out independently. Each of the measurements lasts for 2400 consecutive time slots. The events and the time stamps for UDP flows are tracked by the system.

5.3.2 Numerical Results

Figures 5.5, 5.6, and 5.7 illustrate respectively the flow completion probability, the flow blocking probability, and the flow termination probability as a function of α. α is the parameter in Fig. 5.3 that represents the probability of PUs' re-appearance. When α increases, PUs return more often to the channels and the opportunities for the UDP transmissions will be reduced, hence the probability of flow completion decreases and the blocking probability and forced termination probability increase. Though the probability curves in these figures are not smooth due to the limited number of samples being measured and the stochastic property of the radio channels, we are able to validate the measurement results by confirming

---

[1]To convert the capacity that is defined in the previous chapters in percentage, we can divide the capacity by the arrival rate of the SU flows, i.e., λ_S.

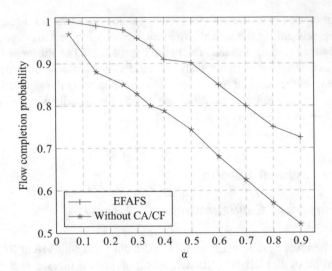

Fig. 5.5 The flow completion probability obtained from the test-bed measurement

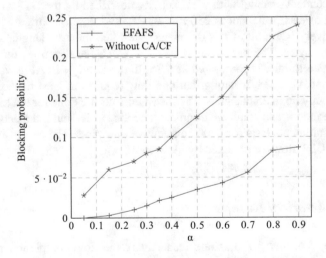

Fig. 5.6 The flow blocking probability obtained from the test-bed measurement

$(1 - P_b)(1 - P_{fd}) = \rho_{pc}$. Take the results of the EFAFS strategy as an example, when $\alpha = 0.5$, the blocking probability P_b, as shown in Fig. 5.6, is 0.035, and according to Fig. 5.7, the forced termination probability P_{fd} is 0.067. Therefore, $(1 - P_b)(1 - P_{fd}) = 0.9042$, which is consistent with the result for EFAFS presented in Fig. 5.5 when $\alpha = 0.5$.

The measurement results confirm the advantages of applying CF and CA in the channel accessing strategy. In the test-bed system that operates in a time-slotted manner, a strategy with CA and CF is also able to increase the bandwidth for a traffic flow and accommodate more flows simultaneously into the system, hence enhance

Fig. 5.7 The forced termination probability obtained from the test-bed measurement

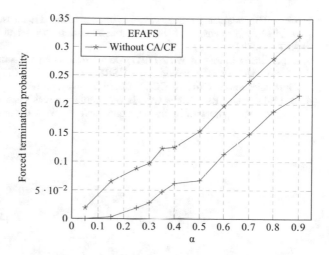

the system performance. Besides, by virtue of the flexibility of the software defined radio platform in channel accessing, the number of channels that are utilized by UDP flows can be adjusted accordingly to the availability of the channel resource, which is another reason that the system performance is improved by the CA and CF enabled strategy.

5.4 Discussions and Summary

In this chapter, we utilized an NI-USRP SDN platform and the LabView software platform to establish a test-bed for transmitting UDP flows on an on-shelf LTE framework. The measurement results indicate the advantages of applying CA and CF in a real system operating in a time-slotted manner for UDP flow communication, and therefore confirm the conclusions from the theoretical study.

The test-bed system for studying the impact of CA and CF can be extended by applying other transport layer protocols than the UDP, such as the TCP and the stream control transmission protocol. One can also employ the channel sensing functionality to make the system a complete CR test-bed platform and see if imperfect channel sensing would affect the final results.

References

1. Jordbru T, Jiao L, Cenkeramaddi LR (2018) UDP flows in cognitive radios with channel aggregation and fragmentation: a test-bed based evaluation. In: IEEE International conference on advanced networks and telecommunications systems (ANTS), Radisson Blu, Indore, India

2. Welch TB, Shearman S (2012) Teaching software defined radio using the USRP and LabVIEW. In: 2012 IEEE international conference on acoustics, speech and signal processing (ICASSP), pp 2789–2792
3. Liu Z, Zhang L, Rao W, Wu Z (2017) Demonstrating high security subcarrier shifting chaotic OFDM cognitive radio system using USRP. In: 2017 IEEE/CIC international conference on communications in China (ICCC), IEEE, Piscataway, pp 1–6
4. NI (2017) NI USRP-2953r – LabVIEW communications system design suite 2.0 – national instruments. http://www.ni.com/documentation/en/labview-comms/latest/2953r/overview/

Printed in the United States
By Bookmasters